Woodhead Publishing in Food Science, Technology and Nutrition

Lipid glossary 2

Frank D. Gunstone

Honorary Professor, Scottish Crop Research Institute,
Dundee, UK

Bengt G. Herslöf

Managing Director, Scotia LipidTeknik AB,
Stockholm, Sweden

WOODHEAD
PUBLISHING

Oxford Cambridge Philadelphia New Delhi

Published by Woodhead Publishing Limited,
80 High Street, Sawston, Cambridge CB22 3HJ, UK
www.woodheadpublishing.com; www.woodheadpublishingonline.com

Woodhead Publishing, 1518 Walnut Street, Suite 1100, Philadelphia,
PA 19102-3406, USA

Woodhead Publishing India Private Limited, G-2, Vardaan House, 7/28 Ansari
Road, Daryaganj, New Delhi – 110002, India
www.woodheadpublishingindia.com

First published by The Oily Press, 2000.
Reprinted by Woodhead Publishing Limited, 2013

© PJ Barnes & Associates, 2000; © Woodhead Publishing Limited, 2012
The authors have asserted their moral rights

British Library Cataloguing in Publication Data
A catalogue record for this book is available from the British Library

ISBN 978-0-9531949-2-6 (print)
ISBN 978-0-85709-797-2 (online)

This book is **Volume 12** in **The Oily Press Lipid Library**

Produced in cooperation with Karlshamns Oils and Fats Academy

Preface to *Lipid Glossary 2*

Our Preface to *A Lipid Glossary* published in 1992 (see page *v*) ended with "if this book is well-received we hope to correct it, update it, and extend it in a few years time...". The book was well received and this was reflected in its high sales.

How does *Lipid Glossary 2* published in 2000 differ from the slimmer volume of 1992?

• We have extended the text by adding new entries, by extending existing entries, and by adding key references to many of them. We have also used more graphics to depict molecular structures. The number of entries has been raised from 900 to over 1200, and graphics from about 60 to over 180. As a consequence, the main text has increased from 100 to 237 pages.

• The book is still replete with cross references but these are now indicated in italic rather than in bold type. This gives a better appearance to the pages of text. However it does mean that some botanical references which are italicized do not have a cross reference.

• We have also revised the Appendices. Appendices 1–3 are deleted but all the information is now contained in the main text. Appendices 4 and 5 have been retained (though renumbered) and updated. No information has been lost through these changes except that we have removed some of the earlier book titles from the list.

We hope this new glossary will be as popular and as useful as the first. We offer it as a handy reference for a wide range of lipid scientists and technologists as well as those involved in the business and commerce of these materials. Many companies bought several copies of *A Lipid Glossary* for wide distribution to their staff. They recognized the importance of having a copy readily available for quick consultation.

As with *A Lipid Glossary*, we thank Bill Christie who has acted as friend, mentor, adviser, and consultant for his generous help and advice. We also thank Peter Barnes for his efforts in connection with this volume. For help with typing and formulas we thank Birgitta Taube, Marie Sedig, Jan Holmbäck and Anders Viberg. Kersti Sundbro gave assistance from the Scotia LipidTeknik library for which we are grateful.

Frank Gunstone (Dundee)

Bengt Herslöf (Stockholm)

March 2000

Preface to *A Lipid Glossary* (published 1992)

Having produced this Lipid Glossary it is incumbent upon us to indicate our understanding of the word lipid. There is no exact, succinct and agreed definition of lipids and whilst there is no disagreement over a wide range of compounds, there is a gray zone of compounds not uniformly accepted as lipids. There are several problems. Are lipids only natural products and do we exclude the wide range of synthetic compounds? We think not. Are lipids adequately defined in terms of their solubility in a range of `fat solvents'? Solubility is linked with structure but it seems a rather vague criterion designed to distinguish typical lipids from typical carbohydrates and proteins and leaves many doubtful cases. We prefer a definition of lipids based mainly, if not entirely, on structure and related, in turn, to the underlying biosynthetic pathways by which fatty acids are produced.

The acetate–malonate pathway of biosynthesis leads, through simple variants, to three major categories of natural products: fatty acids by a reductive pathway, isoprenoids via mevalonate, and phenols by cyclisation of polyketides.

We consider that lipids are compounds based on fatty acids or closely related compounds such as the corresponding alcohols or the sphingosine bases.

We accept, of course, that in some complex compounds the acyl chain is less significant and less important than the polar head group. This is true of some complex glycolipids. We believe further that increasing understanding of biological membranes will justify the importance of the acyl part of even complex lipid molecules. Our definition includes all the major groups of materials generally recognized as lipids. It includes the sterol esters but not the free sterols and we accept this omission. To be interested in lipids, however, does not exclude an interest in compounds which though not strictly lipids are nevertheless closely associated with them such as some sterols and the tocopherols. This definition is not very far away from that of Christie: 'Lipids are fatty acids and their derivatives, and substances related biosynthetically or functionally to these compounds'.

Included in the glossary are the names of fatty acids and lipids, the major oils and fats, terms associated with their analysis, refining, and modification, and the major journals and societies concerned with lipid chemistry.

Lipids have become the concern of a range of disciplines — from medicine to organic chemistry, from food applications to pharmaceutical formulations, from oleochemicals to analytical reference compounds, to mention only a few examples. However, to some extent separate terminologies have developed and terms are sometimes used without an understanding of the original meaning. Many synonyms and trivial names are used and we have felt the need to collect all terms that we think are related to lipids, to explain them and hopefully to clarify the important concepts. We have written this glossary for all those who are working with lipids and who sometimes need a reminder, like ourselves, but especially for those who are entering into the lipid field and find themselves overwhelmed by the many new terms and trivial names they meet.

We see our book as a source of information for those who work with lipids and for those who find themselves needing to understand the terminology of oil/fat/lipid scientists and technologists. The major part of the book is the glossary which contains brief and simple definitions of 900 terms. Many cross references are included to make it easier to find information from different starting points. These are often but not always indicated by words in bold type. But we have also included a number of appendices. Some of these represent a collation of information given in the glossary in more detail. Others represent listings of what we believe to be additional useful information. These appendices are detailed in the Contents and are a valuable addition to the glossary itself.

If this book is well-received we hope to correct it, update it, and extend it in a few years time and we invite readers to contact either of us with comments and suggestions. We wish to thank our publisher (and friend), Bill Christie, for his help and advice in producing this book.

Frank D. Gunstone (St Andrews)

Bengt G. Herslöf (Stockholm)

January 1992

CONTENTS

CONTENTS

A

abietic acid
a major resin acid present in tall oil fatty acids. *Tall oil* is a by-product of the wood pulp industry and consists of almost equal parts of fatty acids (mainly *oleic* and *linoleic*) and a series of cyclic isoprenoid acids like abietic acid along with some triterpenes. The fatty acids and resin acids can be separated by distillation.
A. Hase et al., Tall oil as a fatty acid source, Lipid Technology, 1994, 6, pp.110-114.

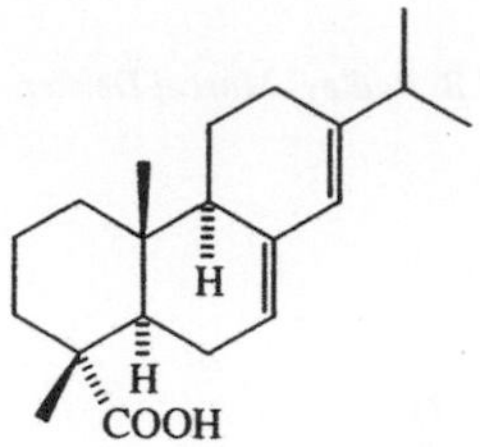

Abietic acid
$C_{20}H_{30}O_2$
Mol. Wt.: 302.5

accelerated oxidation tests
methods of estimating the *induction period* (and hence the shelf life under normal storage conditions) of a fat or fatty food. *Peroxide values* are measured at time intervals whilst the material is held at an elevated temperature (e.g. 100°C). Results obtained at these temperatures must be interpreted with care when using them to predict shelf life since the mechanisms of oxidation change with temperature. See *active oxygen method.*
E.N. Frankel, Lipid Oxidation, The Oily Press, 1998.

acceptable daily intake
defined as the average amount of a food additive that can be safely consumed every day for life. It is expressed as mg of additive per kg of body weight per day.

aceituno oil
oil from the seeds of *Simarouba glauca*. The oil contains palmitic (12%), stearic (28%), and oleic acid (58%), with a low level of linoleic acid (2%).

acetate
salt or ester of *acetic acid.*

Acetems ™
produced commercially by reaction of mono or diacylglycerols with acetic anhydride to give products with 50, 70 or 90% of the free hydroxyl groups acetylated. They have lower melting points than the monoacylglycerols from which they are prepared. Partially acetylated products (50 or 70%) are used in toppings, whipping emulsions and cake mixes. Fully acetylated acetems form flexible films with low penetration of oxygen or water vapour and are used as coating agents for frozen meats or fruit. These compounds are designated E472(a) in Europe and 172,828 in the USA.
N. Krog, Lipid Technologies and Applications (ed. F.D. Gunstone and F.B. Padley) Marcel Dekker, New York (1997), pp.521-534.

1-palmitoyl-2,3-diacetyl-*sn*-glycerol
$C_{23}H_{42}O_6$
Mol. Wt.: 414.6

acetic acid
ethanoic acid (2:0), the fatty acid of lowest molecular weight to occur naturally in lipids. It occurs only rarely but it is the only fatty acid present in *platelet-activating factor*. It is also present as a minor component of milk fat, occurs unusually in some seed oil triacylglycerols (e.g. *Euonymus verrucosus*), and in synthetic monoacylglycerol derivatives (*Acetems*™). It is an important biosynthetic unit — usually as a coenzyme ester such as *acetyl-CoA* — in the production of many natural products including the fatty acids.
J.L. Harwood, Lipid Synthesis and Manufacture (ed. F.D. Gunstone) Sheffield Academic Press, Sheffield (1999) pp.422-466

CH_3COOH

Acetic acid (ethanoic acid, 2:0)
$C_2H_4O_2$
Mol. Wt.: 60.1

acetoglycerides
see *acetylated monoacylglycerols*.

acetolipids
lipids containing *acetic acid*.

acetone-insoluble
This refers to that part of crude commercial *lecithin* which is insoluble in acetone. It is mainly phospholipids. The triacylglycerols are acetone-soluble.

acetone precipitation
laboratory and industrial method to separate triacylglycerols from acetone-insoluble lipids, e.g. phospholipids. This is used on a large scale to produce phospholipids from crude lecithins. See *de-oiling*.
M. Schneider, Industrial production of phospholipids – lecithin processing, Lipid Technology, 1997, 9, pp.109-116. M. Schneider, Lipid Technologies and Applications (ed. F.D. Gunstone and F.B. Padley) Marcel Dekker, New York (1997), p.51. B.F. Szuhaj (ed), Lecithins — sources, manufacture, and uses, AOCS Press, Champaign, USA (1989).

acetyl value
the number of milligrams of potassium hydroxide required to neutralize the acetic acid capable of combining by acetylation with one gram of oil or fat. It is a measure of the content of free hydroxyl groups in the material.

acetylated monoacylglycerols
acetoglycerides, normally produced from monoacylglycerols and acetic anhydride or by *interesterification* of monoacylglycerols with triacetin. The acetylated monoacylglycerols have film-forming properties and are useful in various food applications as coating agents. See *Acetem^{TM}*.

acetylated monoglycerides
see *acetylated monoacylglycerols*.

acetylenic acids
fatty acids with one or more triple bonds (i.e. C≡C). Examples of natural acetylenic acids include *tariric acid* (6a-18:1, mp 50.4-51°C) and *crepenynic acid* (9c12a-18:2). Other acetylenic acids include: *exocarpic, helenynolic, isanic, pyrulic, stearolic, santalbic* (ximenynic), and acids such as 6a9c12c-18:3, 6a9c12c15c-18:4 and 8a11c14c-20:3 which are closely related to the more common polyene acids. It is possible that natural *allenic* and *cyclopropene* acids are made in nature from acetylenic precursors. There

also many natural acetylenic compounds which are not carboxylic acids but are produced biologically from fatty acids, probably via *crepenynic acid*. Oleic acid can be converted to *stearolic acid* (9a-18:1) by chemical procedures. Acetylenic acids are higher melting than their olefinic analogues. They are used as intermediates in the chemical synthesis of long-chain olefinic acids. They can be built up from acetylene (ethyne) and converted into *cis* or *trans* olefinic derivatives by stereospecific partial hydrogenation.

J.-M. Vatèle, Lipid Synthesis and Manufacture (ed. F.D. Gunstone) Sheffield Academic Press, Sheffield (1999) pp.1-45. R.O. Adlof, Lipid Synthesis and Manufacture (ed. F.D. Gunstone) Sheffield Academic Press, Sheffield (1999) pp.46-93. F. Camps and A. Guerrero, Lipid Synthesis and Manufacture (ed. F.D. Gunstone) Sheffield Academic Press, Sheffield (1999) pp.94-126.

acid anhydride

a reactive derivative of *carboxylic acids* used for *acylation*. Anhydrides [$(RCO)_2O$] are made from carboxylic acids by reaction with acetyl chloride or acetic anhydride. They can also be made by reaction with dicyclohexylcarbodiimide at ambient temperature. Though less reactive than *acid chlorides* they are effective acylating agents for OH and NH_2 groups.

F.D. Gunstone, Lipid Synthesis and Manufacture (ed. F.D. Gunstone) Sheffield Academic Press, Sheffield (1999) pp.321-346.

acid chloride

a reactive derivative of *carboxylic acids* used for *acylation*. Acid chlorides (RCOCl) are generally made from carboxylic acids by reaction with phosphorus trichloride, phosphorous pentachloride, phosphorous oxychloride ($POCl_3$), phosgene ($COCl_2$), oxalyl chloride (ClCOCOCl), thionyl chloride ($SOCl_2$) or triphenylphosphine and carbon tetrachloride. Phosphorus trichloride and phosgene are most used on the large scale and oxalyl chloride is useful on a laboratory scale. Acylation of alcohols or amines liberates hydrogen chloride. The presence of a base such as pyridine facilitates the reaction and removes the hydrogen chloride.

F.D. Gunstone, Lipid Synthesis and Manufacture (ed. F.D. Gunstone) Sheffield Academic Press, Sheffield (1999) pp.321-346.

acid oil

a product obtained by the acidification of *soapstock*. It contains appreciable amounts of fatty acid (40–80%) along with glycerol esters and unsaponifiable material.

acidolysis

the formation of esters through reaction of an ester, such as a triacylglycerol, with a fatty acid in the presence of an acidic or enzymic catalyst. The reaction results in exchange of acyl groups. For example it is possible in this way to introduce lauric acid into natural fats containing C_{16} and C_{18} acyl chains and to make long-chain vinyl esters from fatty acid and vinyl acetate. An example of an enzymic process involves the production of *StOSt* (a component of chocolate) from high-oleic sunflower oil (rich in triolein) and stearic acid under the influence of *Mucor miehei* lipase.

acid value
free (unesterified) fatty acid, present in an unrefined oil or remaining in the oil after refining, is measured by titration with alkali and the result is expressed as acid value (mg potassium hydroxide required to neutralise 1 g of fat) or as % of *free fatty acid*. The acid value of a crude oil is reduced during refining by *neutralisation* or *physical refining*. Standard methods are described by AOCS [Cd 3a 63(89)] and IUPAC (2.201).

ACP
see *acyl carrier protein.* .

active oxygen method
AOM or Swift Stability Test. This is a method for measuring the oxidative stability of oils and fats. The induction period is determined from a plot of peroxide value against time when air is bubbled through at $98\pm0.2°C$ and is based on the time required to reach a specific peroxide value (e.g. 100). Standard methods are described by AOCS [Cd 12 57(89)].
E. N. Frankel, Lipid Oxidation, The Oily Press, Dundee (1998).

acylating agent
carboxylic acids or their active derivatives used to acylate OH or NH_2 compounds. The *carboxylic acids* usually require an acidic catalyst (sulphuric acid, hydrogen chloride, boron trifluoride), dicyclocarbodiimide and catalyst (4,4-dimethylaminopyridine, DMAP) or an enzyme. The more reactive *anhydrides* [$(RCO)_2O$] or *acid chlorides* (RCOCl) do not require a catalyst. Thiol esters act as acylating agents in biochemical systems.
F.D. Gunstone, Lipid Synthesis and Manufacture (ed. F.D. Gunstone) Sheffield Academic Press, Sheffield (1999) pp.321-346.

acylation
the replacement of hydrogen, usually in an OH or NH group, by an *acyl group*. The reaction with OH is an *esterification* process and involves reaction with an acid (and catalyst), *acid chloride*, or *acid anhydride*. This is an important step in the chemical synthesis of glycerol esters and also in their biosynthesis

when the acylating agent is usually an *acyl-CoA* derivative operating under the influence of an *acyl transferase*.

P.E. Sonnet, Lipid Synthesis and Manufacture (ed. F.D. Gunstone) Sheffield Academic Press, Sheffield (1999) pp.162-184. R. Bittman, Lipid Synthesis and Manufacture (ed. F.D. Gunstone) Sheffield Academic Press, Sheffield (1999) pp.185-207. F.D. Gunstone, Lipid Synthesis and Manufacture (ed. F.D. Gunstone) Sheffield Academic Press, Sheffield (1999) pp.321-346. J.L. Harwood, Lipid Synthesis and Manufacture (ed F.D. Gunstone) Sheffield Academic Press, Sheffield (1999) pp.422-466.

acyl carnitine
see *carnitine*.

acyl carrier protein
ACP; carrier of acyl groups in the biosynthesis of fatty acids. This protein (9000 Daltons) is central to the fatty acid synthase complex and binds covalently to the acyl intermediates. The acyl derivative is often represented as acyl-ACP. See also *acyl coenzyme A*.

J.L. Harwood, Lipid Synthesis and Manufacture (ed. F.D. Gunstone) Sheffield Academic Press, Sheffield (1999) pp.422-466.

acyl coenzyme A
the activated form of fatty acids required for many biochemical reactions. The acyl group is linked to the coenzyme, a complex nucleotide, as a thiol ester. The acyl derivative is often referred to as acyl-CoA. See also acyl carrier protein.

J.L. Harwood, Lipid Synthesis and Manufacture (ed. F.D. Gunstone) Sheffield Academic Press, Sheffield (1999) pp.422-466.

Acyl Coenzyme-A

acylglycerol
systematic name for all types and combinations of fatty acids esterified to glycerol, e.g. *mono*, *di-* and *triacylglycerols*, often called mono-, di- and triglycerides.

acylglycoses
glycolipids consisting of a mono- or oligo-saccharide partially esterified by fatty acids. These occur naturally and are also produced on an industrial scale for use as *surfactants*.

acyl group
a fatty acid residue (RCO), present in acids (RCOOH), esters (RCOOR'), and amides (RCONH$_2$).

acyl migration
the process by which chemically-bonded acyl groups move to adjacent OH (or NH$_2$) groups. This may occur spontaneously in mono and diacylglycerols. It is particularly important to avoid acyl migration during the synthesis of mixed glycerol esters. This is achieved by using blocking groups and by avoiding experimental conditions, such as high acidity or high alkalinity, which favour acyl migration. See also *interesterification* and *transesterification*.
P.E. Sonnet, Lipid Synthesis and Manufacture (ed. F.D. Gunstone) Sheffield Academic Press, Sheffield (1999) pp.162-184.

acyl transferases
enzymes which promote the acylation of glycerol and related molecules in the biosynthesis of *triacylglycerols* and *phospholipids*. They are generally stereospecific, reacting with a particular hydroxyl group such as the sn–1, sn–2 or sn–3 positions of glycerol or acylated glycerols.
J.L. Harwood, Lipid Synthesis and Manufacture (ed. F.D. Gunstone) Sheffield Academic Press, Sheffield (1999) pp.422-466.

ADI
see *acceptable daily intake*.

adipic acid
the C$_6$ dibasic acid (hexanedioic, mp 151°C). See *dibasic acids*. It is an important industrial chemical made from cyclohexane and is used in the

production of polyamides such as Nylon™-66 and of polyesters. It could also be made by *ozonolysis* of *petroselinic acid*.

Adipic acid
$C_6H_{10}O_4$
Mol. Wt.: 146.1

adipocytes
fat cells. These are specialized cells in the adipose tissue of animals in which most of the cell volume is filled with triacylglycerol in the form of oil droplets. The triacylglycerols are mobilised to meet the energy requirements of the animal when required.

adipose tissue
consists of *adipocytes* and is widely distributed in the mammal body (e.g. under the skin, in the abdominal cavity, around the deep blood vessels). It serves as an energy reserve and insulation medium.

D.R. Body, The lipid composition of adipose tissue, Progress in Lipid Research, 1988, 27, 1, 39-60.
K.N. Seidelin, Fatty acids composition of adipose tissue in humans. Implications for the dietary fat–serum cholesterol CHD issue, Progress in Lipid Research, 1995, 34,3, 199-217.

adrenic acid
a trivial name for the 22:4 (*n–6*) acid 7,10,13,16-*docosatetraenoic acid*. This is one of the less common members of the *n–6 family* of polyene acids resulting from *chain-elongation* of arachidonic acid (*eicosatetraenoic acid*). It is a major component in adrenal lipids. See *n–6 family*.

Adrenic acid
all *cis*-7,10,13,16-docosatetraenoic acid
(7,10,13,16-22:4)
$C_{22}H_{36}O_2$
Mol. Wt.: 332.5

adsorbents
see *adsorption chromatography*.

adsorption chromatography
chromatography, normally planar (thin-layer chromatography) or column, based on adsorbents such as *silica* or *alumina* as the stationary phase. The mobile phase is generally non-polar for non-polar lipids (e.g. hexane with a small proportion of diethyl ether) and polar for the polar lipids (e.g. isopropanol/water mixtures).

Advances in Lipid Research
periodic series of reviews; published by Academic Press since 1963.

aerobic desaturation
the most common biosynthetic reaction leading to the introduction of a double bond. The process requires oxygen and a specific enzyme with a di-iron unit at the active site. This reaction is involved in the conversion of saturated to *monoene* acids and of monoene to *polyene* acids. For example *stearate* is converted to *oleate* and then, in the plant kingdom only, oleate is converted to *linoleate* and *linolenate*. Chain extension and further desaturation of oleate, linoleate, and linolenate lead to the *n*–9, *n*–6, and *n*–3 *families* of polyene acids.
J.L. Harwood, Lipid Synthesis and Manufacture (ed. F.D. Gunstone) Sheffield Academic Press, Sheffield (1999) pp.422-466.

AFECG
L'Association Française pour l'Etude des Corps Gras. A French professional organisation concerned with lipids. It is responsible for the publication of the journal *Oleagineux Corps gras Lipides.*

aflatoxins
fungal toxins (mycotoxins) which frequently contaminate tropical oilseeds (especially peanut and cotton) and the oil and meal derived from them. Some aflatoxins are highly toxic (carcinogenic, mutagenic, teratogenic) but they can be removed by careful refining. The maximum permitted concentration is in the range 1–10 ppb depending on the type of product.

alchornoic acid
a natural epoxy acid. This is *cis*-14,15-epoxy-*cis*-11-eicosenoic acid which occurs naturally in *Alchornea cordifolia* seed oil. It is

the C_{20} homologue of the more widely occurring C_{18} fatty acid acid, *vernolic*.

Alchornoic acid
14*S*-15*R*-Epoxy-11*Z*-eicosenoic acid
$C_{20}H_{36}O_3$
Mol. Wt.: 324.5

alcohols
see *fatty alcohols*.

alcoholysis
a method of making esters involving reaction of another ester (e.g. a triacylglycerol) with an alcohol in the presence of acidic (sulphuric acid, hydrogen chloride, boron trifluoride) or alkaline (sodium methoxide) catalysts or appropriate enzymes. The reaction results in an exchange of the alcohol moiety in esters. Examples include *methanolysis* and *glycerolysis*.

$$RCOOR^1 + R^2OH \rightarrow RCOOR^2 + R^1OH$$

aldehydes
see *fatty aldehydes*.

aldehyde value
a measure of the aldehydes formed by oxidative cleavage of the unsaturated acyl groups in oils and fats. It has never been officially adopted. See also *anisidine value*.

alepramic, aleprestic, alepric, aleprolic, aleprylic acids
see *cyclopentenyl acids*.

Aleurites species
Aleurites fordii and *A. montana* are the source of *tung oil* (China wood oil) which is characterised by its high level of *α-eleostearic acid*.

aleuritic acid
9,10,16-trihydroxypalmitic acid (mp 102°C). This is a tri*hydroxy acid*

produced from shellac resin by hydrolysis. The natural acid is the (±)-*threo* isomer.

alkali fusion
reaction with alkali (sodium or potassium hydroxide) at elevated temperatures (180–270°C). The reaction is used on a commercial scale to convert *castor oil* or *ricinoleic acid* to *2-octanol* and *sebacic acid.*

alkali isomerization
the movement of one or more double bonds under the influence of alkali. For example, potassium hydroxide in ethanediol at 170 or 180°C converts *linoleate* mainly to a mixture of the 9*c*11*t* and 10*t*12*c* isomers (both *conjugated*) and linolenate to a more complex mixture containing both conjugated dienes and trienes. These can be measured quantitatively at 234 nm (dienes) and 268 nm (trienes) by *ultraviolet spectroscopy.* This method was used to measure linoleate and linolenate in vegetable oils before the general use of *gas chromatography.* It is now being used without solvent to prepare *CLA* commercially from linoleic-rich oils such as *sunflower* or *safflower.*

alkali refining
one of the refining processes to which crude oil is submitted. It usually follows *degumming* and precedes *bleaching.* It is designed to remove free *fatty acid* and involves treatment with alkali. *Soap* is removed by centrifugation and neutralised oil is washed with water. This process is also called neutralisation.
D.A. Allen, Lipid Technologies and Applications (ed. F.D. Gunstone and F.B. Padley) Marcel Dekker, New York (1997), pp.137-167. W. De Greyt and M. Kellens, Edible Oil Processing (eds W. Hamm, R.J. Hamilton) Sheffield Acdemic Press, Sheffield (2000), pp.90-94.

alkanoic acids
the general name for saturated aliphatic carboxylic acids. The straight-chain members with the general formula $CH_3(CH_2)_n COOH$ are the best known.

These occur in nature, mainly from C_2 (ethanoic, *acetic*) to C_{30} and above The most common are the C_{12} (lauric, *dodecanoic*), C_{14} (myristic, *tetradecanoic*), C_{16} (palmitic, *hexadecanoic*), and C_{18} members (stearic, *octadecanoic*). Other examples are cited under their own names.

alkenylglycerols
see *plasmalogens*.

alkylglucosides
mono alkyl ethers of glucose (and glucose oligomers) made commercially from glucose and fatty *alcohols* (both are natural products). The compounds have valuable surfactant properties and are subject to increasing demand.
K. Hill et al. (ed), Alkyl polyglycosides – Technology, Properties and Applications, VCH, Weinheim, Germany (1997).

alkylglycerols
see *ether lipids*.

alkyl ketene dimer
a reactive acylating agent used for sizing paper with which it reacts both chemically and physically. It is made, for example, from hydrogenated *tallow*, via the *acid chloride*.

Alkyl ketene dimer

allenic acids
natural fatty acids with allenic double bonds, i.e. with the $-C=C=C-$ group. These acids are chiral by virtue of the allenic group. Examples are laballenic acid (1) and lamenallenic acid (2):

$$H_3C(CH_2)_{10}CH=C=CH(CH_2)_3COOH \qquad (1)$$
$$H_3CCH=CH(CH_2)_8CH=C=CH(CH_2)_3COOH \qquad (2)$$

A C_8 hydroxy allenic acid (8-OH 2,3-8:2 *stillingic*) is present in *stillingia oil* (8–10%) where it is associated with 2t4c-10:2 to make a C_{18} estolide. Many C_{11} allenic acids have been synthesised.

almond oil
nut oil from *Prunus amygdalus*. It is rich in oleic acid (>60%) and is used in cosmetic and pharmaceutical formulations. It is listed in pharmacopoeia. See also *speciality oils*.

alumina
Al_2O_3. Often used as a chromatographic adsorbent. See *adsorption chromatography*.

ambrettolic acid
this acid, present in musk seed oil, is 16-hydroxy-7-hexadecenoic acid. It melts at 53–55°C (α form) or 26–27°C (β form). Its lactone (7-hexadecan-16-olide) exists in *cis* and *trans* forms both of which are used in the perfume industry.

Ambrettolic acid
16-hydroxy-7-hexadecenoic acid
(16-OH-7-16:1)
$C_{16}H_{30}O_3$
Mol. Wt.: 270.4

American Oil Chemists' Society
founded 1921. Publishes four journals: *Journal of the American Oil Chemists' Society* (formerly under other names) since 1923, *Lipids* since 1966, *INFORM* (International News on Fats, Oils and Related Materials) since 1990 and *Journal of Surfactants and Detergents* from 1998. The society is based in Champaign, Illinois, USA.

AMF
see *anhydrous milk fat*.

amine oxides
compounds made from tertiary amines by reaction with hydrogen peroxide. They have structures such as those shown on the next page in which one or two of the R groups are long chain. They are important *surface-active compounds*.

G. Bognolo, Lipid Technologies and Applications (ed. F.D. Gunstone and F.B. Padley) Marcel Dekker, New York (1997), p.660.

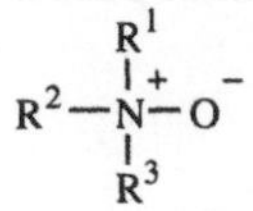

$$R^2-\overset{\displaystyle R^1}{\underset{\displaystyle R^3}{\overset{|}{\underset{|}{N}}}}\overset{+}{-}O^-$$

Amine oxide

amines

produced from fatty acids by the *oleochemical industry* for use as surface active compounds. They are made from fatty acids via amide and nitrile:

fatty acid → amide → nitrile → primary and secondary amine

The primary and secondary amines can be converted to *tertiary amines*, *quats*, and *amine oxides*, all of which have valuable surface-active properties.

A.D. James, Lipid Technologies and Applications (ed. F.D. Gunstone and F.B. Padley) Marcel Dekker, New York (1997), pp.609-631.

amphiphilic

practically all lipids are amphiphilic by reason of the fact that they have a hydrophilic part, e.g. a polar head group, and one or more *lipophilic* carbon tails, e.g. long-chain acyl groups. Important amphiphiles are the *phosphatidylcholines* and other *polar lipids*.

anaerobic pathway

The more common biosynthetic route to unsaturated acids is *aerobic desaturation*, an oxygen-dependent process. However microorganisms in an anaerobic environment produce unsaturated acids by an alternative non-oxygen-dependent pathway which is a modification of the *de novo pathway*. For example, 9c-16:1(*palmitoleic acid*) and 11c-18:1 (*vaccenic acid*) are produced as shown below. 2-Hydroxydecanoate is dehydrated to the 3c isomer (rather than the more usual 2t acid) and then chain-extended to 16:1 and 18:1 acids.

2-OH-10:0 → 3c-10:1 → 5c-12:1 → 7c-14:1 → 9c-16:1 → 11c-18:1

J.L. Harwood, Lipid Synthesis and Manufacture (ed. F.D. Gunstone) Sheffield Academic Press, Sheffield (1999) pp.422-466.

anandamide
arachidonyl ethanolamide, an endogenous eicosanoid that binds specifically to the cannabinoid receptor in the brain.

Anandamide
$C_{22}H_{37}NO_2$
Mol. Wt.: 347.5

anchoveta oil
another name for *anchovy oil.*

anchovy oil
major fish oil, which originates from the west coast of South America. One tonne of anchovy will produce 210 kg of meal (still containing 10 % of oil) and 30 kg of oil. These contain ca 7.5 kg of *eicosapentaenoic acid* (EPA) and *docosahexaenoic acid* (DHA) in the oil and 3.5 kg in the meal.

anhydrous milk fat
cow milk fat free of aqueous phase. It is fractionated to give harder and softer fractions which can be blended to give more spreadable butters and are used for a variety of other edible purposes.
E. Flack, Lipid Technologies and Applications (ed. F.D. Gunstone and F.B. Padley) Marcel Dekker, New York (1997), pp.305-327. W. de Greyt et al., Food and non-food applications of milk fat, Lipid Technology, 1993, 5, pp.138-140; Lipase-catalysed modification of milk fat, Lipid Technology, 1995, 7, pp.10-12.

animal oils and fats
a general term for commercial products from animal sources such as *fish oils*, *lard*, *beef tallow* and *mutton tallow*. *Butter* and other dairy products also belong to this category.
J.A. Love, Bailey's Industrial Oil and Fat Products, (ed. Y. H. Hui) John Wiley & Sons, New York (1996), Volume 1, pp.1-18.

anionic surfactants
the most widely used type of surface-active compounds. They contain one acyl or alkyl chain with a negatively charged head group. Common examples include *soaps* (salts of carboxylic acids) and salts of sulphonic acids or sulphates (see formulas on next page).

M.R. Porter, Lipid Technologies and Applications (ed. F.D. Gunstone and F.B. Padley) Marcel Dekker, New York (1997), pp.579-608.

RCOONa   RCH₂OSO₃Na

Anionic surfactants

anisidine value

a measure of secondary oxidation products. These are mainly aldehydes produced by breakdown of *hydroperoxides*. The value is based on assessment of the chromophore at 350 nm produced by reaction of 4-methoxyaniline (anisidine) and aldehydes, especially 2-enals.

E.N. Frankel, Lipid Oxidation, The Oily Press, Dundee (1998).

Typical coloured product formed in measuring the anisidine value

anteiso acids

general name for fatty acids with a branched methyl group on the ω-3 carbon atom which thus becomes a chiral centre. Natural anteiso acids generally have an odd number of carbon atoms in each molecule since they are biosynthesised by chain-elongation of the C_5 acid 2-methylbutanoic, itself a protein metabolite. See also *iso acids*.

Anteiso acid

antioxidants

materials which inhibit *autoxidation*. These may be natural or synthetic and may operate as chain-breakers or by inhibiting the chain initiation step (by chelating metal ions). Antioxidants do not prevent oxidation: they extend the *induction period* during which oxidation is slow and of no great consequence. The chain breaking antioxidants are usually phenols or amines or polyunsaturated compounds. See also *butylated hydroxy anisole, butylated hydroxy toluene, tertbutyl hydroquinone, propyl gallate, ethoxyquin,* and *tocols*.

E.N. Frankel, Natural and biological antioxidants in foods and biological systems, Lipid Technology, 1995, 7, pp.77-80.   J. Loliger, Natural antioxidants, Lipid Technology, 1991, 3, pp.58-61. D.E. Pratt, Bailey's Industrial Oil and Fat Products, (ed. Y. H. Hui) John Wiley & Sons, New York (1996), Volume 3, pp.523-545. E.A. Decker, Food Lipids: Chemistry, Nutrition, and Biotechnology, (ed. C.C. Akoh and D.B. Min) Marcel Dekker, New York (1998) pp.397-421. D. Reische, D.A. Lillard, et al., Food Lipids: Chemistry, Nutrition, and Biotechnology, (ed. C.C. Akoh and D.B. Min) Marcel Dekker, New York (1998) pp.423-448. E.N. Frankel, Lipid Oxidation, The Oily Press, 1998. T. Netscher, Lipid Synthesis and Manufacture (ed F.D. Gunstone) Sheffield Academic Press, Sheffield (1999) pp.250-267. F.C. Thyrion, Lipid Synthesis and Manufacture (ed. F.D. Gunstone) Sheffield Academic Press, Sheffield (1999) pp.268-287.

AOCS
see *American Oil Chemists' Society.*

AOM
see *active oxygen method.*

APAG
Association de Producteurs d'Acides Gras. A trade association of fatty acid producers.

Appetizer shortening
a blend of animal and vegetable fats designed to meet certain nutritional requirements without sacrificing functionality and consumer-appeal for use in *baking, frying* and in *margarines.* It is mainly animal fat (*beef, lard, milk fat*) from which *cholesterol* has been largely removed by steam distillation along with vegetable oil such as *corn oil* which is rich in *linoleic acid.* The final mixture has at least twice as much *linoleic* as *myristic acid* to counteract the cholesterol-raising influence of the latter.

R.D. Kiley, et al., Advances in structured fats: Appetize shortening, Lipid Technology, 1996, 8, pp.5-10.

apricot oil
an oleic-rich oil (ca 60%) from *Prunus armeniaes.* See also *speciality oils.*

arachide (huile d')
French name for *groundnut oil.*

arachidic acid
trivial name for *eicosanoic acid* (20:0, mp 76°C).

Arachis hypogea (Leguminosae)
see *groundnut oil.*

arachis oil
pharmacopoeial term for *groundnut oil*.

arachidonic acid
trivial name for 5c8c11c14c-*eicosatetraenoic acid* (20:4). See *eicosatetraenoic acid*. See also *Arasco*™.

Arasco™
a triacylglycerol oil produced from the fungus *Mortierella alpina* by Martek Biosciences Corporation as a source of arachidonic acid (~40%) for nutritional purposes, especially *infant formula*. Similar products are available from other companies.
D.J. Kyle, Production and use of a single cell oil highly enriched in arachidonic acid, Lipid Technology, 1997, 9, pp.116-120.

argentation chromatography
the use of silver ion as complexing agent for double bonds. Normally, silver nitrate present in the stationary phase retards the migration of unsaturated lipids. See also *silver ion chromatography*.
B. Nikolova-Damyanova, Silver ion chromatography and lipids, Advances in Methodology – One (ed W.W. Christie) The Oily Press, Ayr, (1992) pp.181-237.

artemesic acid
another name for *coriolic acid*.

arteriosclerosis
the general term for thickening and hardening of arterial walls, which may lead to coronary heart disease. See also *atherosclerosis*.

ASA
American Soybean Association, devoted to the promotion of soybean (bean, oil and meal). It is based in St Louis (Missouri, USA) with centres in many other countries.

asclepic acid
another name for *cis-vaccenic acid*, present in *Asclepia* oils.

ascorbic acid
ascorbic acid (vitamin C) is a useful water-soluble antioxidant. To increase its fat solubility it is often used in the form of its palmitic ester. E-numbers are

304 (ascorbyl palmitate), 300 (L-ascorbic acid), 301 (sodium ascorbate), and 302 (calcium ascorbate).

E.N. Frankel, Lipid Oxidation, The Oily Press, Dundee (1998).

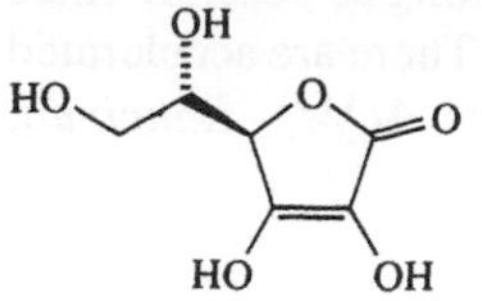

L-Ascorbic acid
$C_6H_8O_6$
Mol. Wt.: 176.1

Aspergillus niger
see *lipases*.

Association Francaise pour l'Etude des Corps Gras
A French group concentrating on the scientific and technical aspects of lipids. Was the publisher of *Revue Francaise des Corps Gras* but this was replaced in 1994 by *Oleagineux Corps gras Lipides*.

atherosclerosis
formation and accumulation of lipid-rich *plaques* which may block arteries and lead to strokes and to coronary heart disease. See also *arteriosclerosis*.

auricolic acid
a C_{20} diene hydroxy acid present in the seed lipid of *Lesquerella auriculata*. It is a homologue of *densipolic acid*.

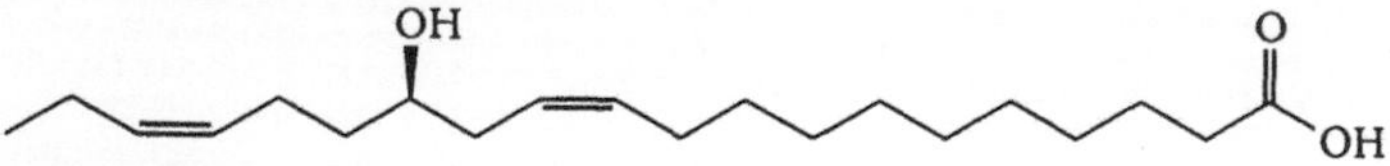

Auricolic acid
14*R*-Hydroxy-11Z,17Z-eicosadienoic acid
$C_{20}H_{36}O_3$
Mol. Wt.: 324.5

autoxidation
the most common form of oxidative deterioration. Reaction occurs by a radical chain process and olefinic *hydroperoxides* are formed. It is promoted by trace metals (especially copper and iron) which act as *pro-oxidants* and is hindered by a range of materials which serve as *antioxidants*. Its progress is

followed by determination of the *peroxide value*. Reaction is slow until there is a build up of reactive species which promote further reaction and then proceeds more rapidly. The time taken to arrive at this quicker stage is the *induction period*. In a fat or fatty food this should be as long as possible since the material becomes *rancid* around the end of this time. There are accelerated methods of assessing the induction period (see *AOM, Rancimat, Oxidograph*).

E.N. Frankel, Lipid Oxidation, The Oily Press, Dundee (1998).

avenasterol

plant sterol, biosynthetic precursor to stigmasterol and sitosterol, occurs as the Δ5 (shown) and Δ7 isomers.

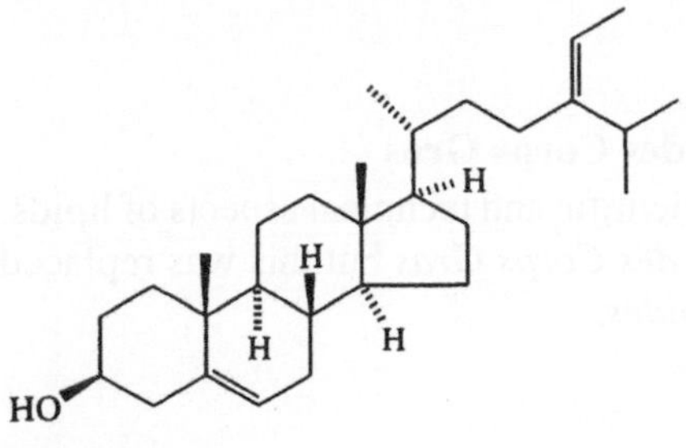

Δ⁵-Avenasterol
$C_{29}H_{48}O$
Mol. Wt.: 412.7

avocado oil

an oleic-rich oil (ca 70%) from the fruit pulp of *Persea americana*. See also *speciality oils*.

azelaic acid

the C_9 dibasic acid (nonanedioic, mp 106°C). This dibasic acid is produced, along with *nonanoic acid*, by *ozonolysis* of oleic and other Δ-9 acids. See also dibasic acids.

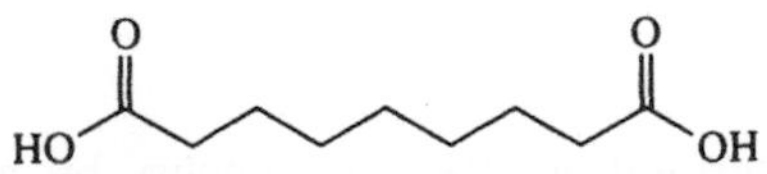

Azelaic acid
nonanedioic acid
$C_9H_{16}O_4$
Mol. Wt.: 188.2

B

bases (long-chain)
see *sphingoid bases.*

***Bassia* species**
see *mowrah butter.*

batyl alcohol
an *sn*–1 alkyl glycerol (*ether lipid*) having the structure shown and melting at 71°C. It occurs, for example, in some fish oils with the two free hydroxyl groups in acylated form.

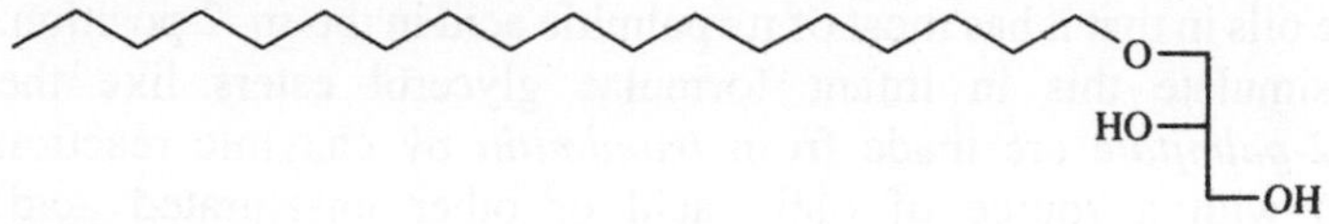

Batyl alcohol
1-octadecyl-*sn*-glycerol
$C_{21}H_{44}O_3$
Mol. Wt.: 344.6

beef tallow
see *tallow.*

beeswax
wax produced by bees from the genus *Apis*. It consists mainly of wax esters (70–80%), free acids (10–15%) and hydrocarbons (10–20%). The esters are largely C_{40}–C_{48} compounds based on palmitic and stearic acids and their ω2 and ω3 hydroxy derivatives. Annual production is about 6700 tonnes. Beeswax is used in candles, packaging, polishes, printing inks, and cosmetics.

R.J. Hamilton, Waxes: Chemistry, Molecular Biology and Functions, (ed. R.J. Hamilton) Oily Press, Dundee (1995) pp.263-264.

behenic acid
trivial name for *docosanoic acid.*

Benecol™
a margarine enriched with sitostanol (a hydrogenated *phytosterol*) usually as

an acyl ester which inhibits *cholesterol* absorption. Sitostanol is obtained as a by-product from the wood pulp industry.

S. von Hellens, Benecol margarine enriched with stanol esters, Lipid Technology, 199, 11, 29-31.

bentonite
see *bleaching earths*.

beta oxidation
see *oxidation (β)*

Betapol ™
a glycerol ester with palmitic acid in the *sn*–2 position produced as a constituent of human milk replacements. *Human milk fat* differs from vegetable oils in that it has most of its palmitic acid in the sn–2 position. In order to simulate this in infant formulae glycerol esters like the *1, 3-dioleate-2-palmitate* are made from *tripalmitin* by enzymic reaction (1, 3-specific) with a source of oleic acid or other unsaturated acid. Palmitic is exchanged in the *sn*–1 and 3 positions but retained in the *sn*–2 position.

G.P. McNeill, Lipid Synthesis and Manufacture (ed. F.D. Gunstone) Sheffield Academic Press, Sheffield (1999) pp.288-320. R.G. Jensen, Human milk lipids as a model for infant formulas, Lipid Technology, 1998, 10, 34-38.

BHA
see *butylated hydroxy anisole*.

BHT
see *butylated hydroxy toluene*.

bilayer
normally refers to the lamellar arrangement of polar lipids in water. The acyl chains form the internal hydrophobic part and the polar head groups the hydrophilic part of the bilayer. Biological *membranes* have this basic structure.

K. Larsson, Lipids — Molecular Organisation, Physical Functions and Technical Applications, The Oily Press, Dundee (1994)

bile acids
primary bile acids (for example, cholic acid as shown in the accompanying structure) are synthesised from *cholesterol* in the liver and transformed by

intestinal bacteria to secondary bile acids (for example, deoxycholic acid). See also the entry for *bile salts.*

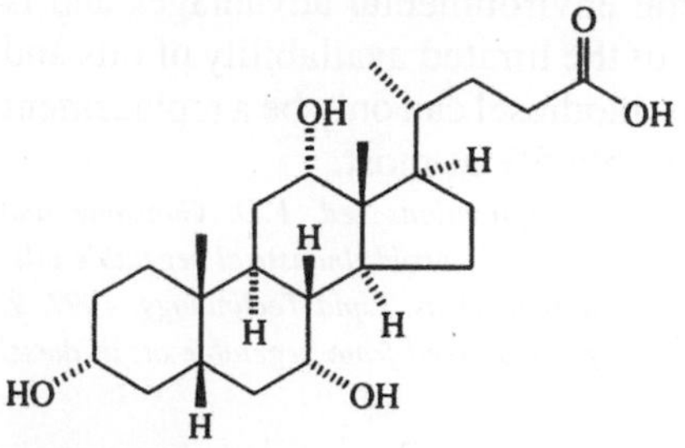

Cholic acid
$C_{24}H_{40}O_5$
Mol. Wt.: 408.6

bile salts

salts of cholic acid and other *bile acids,* e.g. sodium glycocholate and sodium taurocholate. They participate in the emulsification and digestion of lipids in the small intestine.

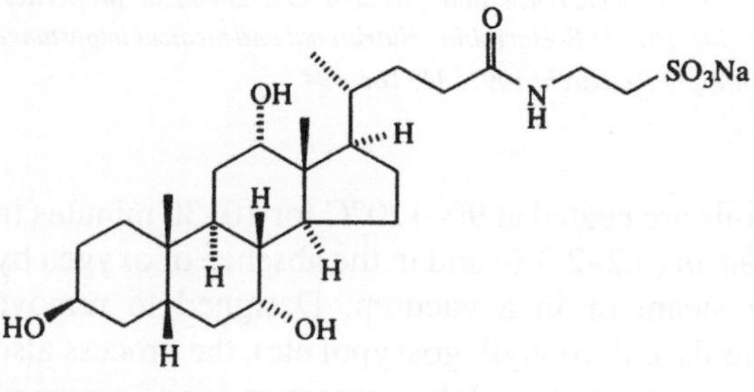

Sodium taurocholate
$C_{26}H_{44}NNaO_7S$
Mol. Wt.: 537.7

biodiesel

the name given to methyl (or other alkyl ester) of long-chain acids produced from an oil or fat for use as an alternative to automotive diesel fuel or as heating fuel. These are most often the methyl esters of *rapeseed oil* (Europe), *soybean oil* (USA), or *palm oil* (Malaysia) but other oils and fats can be used including waste products. The methyl esters are made by *methanolysis*. Alternative names include *biofuel, diester* (diesel ester), *RME,* (rape methyl esters), *soy diesel, palm diesel* etc. Biodiesel is more expensive than hydrocarbon diesel fuel but it has several environmental advantages and is based on a renewable resource. Because of the limited availability of oils and fats compared with the hydrocarbon fuels, biodiesel can only be a replacement for a limited amount of diesel fuel – probably 5% at most.

Y.M. Choo, A.N. Ma, et al., Lipid Technologies and Applications (ed. F.D. Gunstone and F.B. Padley) Marcel Dekker, New York (1997), pp.771-785. S. Harold, Industrial vegetable oils: Opportunities within the European biodiesel and lubricants markets, Lipid Technology, 1997, 9, 33-38 and 67-70. H. Sadeghi-Jorabchi, et al., Use of fuels derived from vegetable oil in diesel engines, Lipid Technology, 1995, 7, 107-111.

biohydrogenation
see *rumen biohydrogenation.*

blackcurrant seed oil
oil obtained from *Ribes niger.* The oil is used as a dietary supplement because it contains γ-linolenic acid (~17%) along with palmitic (7), stearic (4), oleic (11), linoleic (47), α-linolenic (13), and stearidonic acid (3%). Other convenient sources of γ-linolenic acid are *borage oil* and *evening primrose oil.*

F.D. Gunstone, Gamma linolenic acid — occurrence and physical and chemical properties, Progress in Lipid Research, 1992, 31, 145-161. D.F. Horrobin, Nutritional and medical importance of gamma-linolenic acid, Progress in Lipid Research, 1992, 31, 163-194.

bleaching
a refining process in which oils are heated at 90–120°C for 10–30 minutes in the presence of a *bleaching earth* (0.2–2.0%) and in the absence of oxygen by operating with nitrogen or steam or in a vacuum. Designed to remove unwanted pigments (carotenoids, chlorophyll, gossypol etc), the process also removes oxidation products, trace metals, sulphur compounds and traces of soap. Palm oil can be heat-bleached at high temperatures in the absence of bleaching earth.

D.A. Allen, Lipid Technologies and Applications (ed. F.D. Gunstone and F.B. Padley) Marcel Dekker, New York (1997), pp.137-167.

bleaching earth
used in *bleaching* to remove pigments and other impurities from crude oils. The bleaching earths are generally acid-washed clays such as *bentonite* or *Fullers earth*, to which a little activated carbon (5–10%) may be added.

Bligh and Dyer extraction

A procedure developed to extract lipid from animal tissues with the minimum practical volume of organic solvents. Tissues are homogenised with a defined volume of chloroform and methanol and the organic extract subsequently shaken with aqueous potassium chloride solution.

W.W. Christie, Advances in Lipid Methodology — Two (ed. W.W. Christie). Oily Press, Dundee, 1993, pp.195-213.

bloom

the appearance of white spots (fat crystals) on the surface of *chocolate*. Cocoa butter triacylglycerols in chocolate may undergo polymorphic transitions which are the cause of the visible phenomenon. The development of bloom is considered undesirable and compounds such as *Bohenin™* are used to delay the onset of bloom.

F.B Padley, Lipid Technologies and Applications (ed. F.D. Gunstone and F.B. Padley) Marcel Dekker, New York (1997), pp.391-432.

BOB

see *Bohenin™*

Bohenin ™

glycerol 1,3-dibehenate 2-oleate (BOB) which inhibits *fat bloom* when added to *chocolate*. It can be produced by *interesterification* of *triolein* and *behenic* acid or ester (22:0) in the presence of a 1,3-*stereospecific lipase*.

F.B Padley, Lipid Technologies and Applications (ed. F.D. Gunstone and F.B. Padley) Marcel Dekker, New York (1997), pp.391-432.

bolekic acid

an unsaturated C_{18} acid with acetylenic unsaturation, present in a few oils such as isano oil (*Onguekoa gore*).

Bolekic acid
13Z-Octadecene-9,11-diynoic acid
$C_{18}H_{26}O_2$
Mol. Wt.: 274.4

borage seed oil
oil obtained from *Borago officinalis*. Also called starflower oil. The oil is used as a dietary supplement because it contains *γ-linolenic acid* (20-25%). Also present are palmitic (10), oleic (16), linoleic (38), and the C_{20}, C_{22}, and C_{24} monoenes (about 9% total). Other convenient sources of γ-linolenic acid are *blackcurrant seed oil* and *evening primrose oil*. The level of γ-linolenic acid in borage oil can be raised to 40–45% by *enzymic enhancement*.

F.D. Gunstone, Gamma linolenic acid — occurrence and physical and chemical properties, Progress in Lipid Research, 1992, 31, 145-161. D.F. Horrobin, Nutritional and medical importance of gamma-linolenic acid, Progress in Lipid Research, 1992, 31, 163-194.

Borago officinalis
see *borage oil*.

borate-TLC
boric acid or sodium borate is used as a modifier of silica for chromatography to prevent *acyl migration* in *mono-* and *diacylglycerols*. It is also used to separate stereo- and/or regioisomers of hydroxy lipids.

Borneo tallow
see *Shorea robusta*.

bourrache (huile de)
French name for *borage* oil.

branched-chain acids
branched-chain acids, as opposed to straight-chain acids, have a branched carbon skeleton. Most often the branch is one or more methyl groups. See, for example, *iso* and *anteiso acids*, *phytanic* and *pristanic* acids, and *isostearic* acid. Branched-chain acids have lower melting points than their straight-chain analogues and, for this reason, are useful components of lubricants and cosmetics.

Brassica
the genus *Brassica* belongs to the Cruciferae family and includes important plants such as *rape* and *mustard*.

Brassica alba (Cruciferae)
the source of white *mustard seed oil*.

Brassica campestris (Cruciferae)
one source of *rapeseed oil*, see *Brassica napus*.

Brassica hirta (Cruciferae)
one source of yellow *mustard seed oil*.

Brassica napus (Cruciferae)
one source of *rapeseed oil*, see *Brassica campestris*.

Brassica nigra (Cruciferae)
the source of black *mustard seed oil*.

brassicasterol
this sterol is present in rape and mustard seed oils and in some other seed oils at a much lower level.

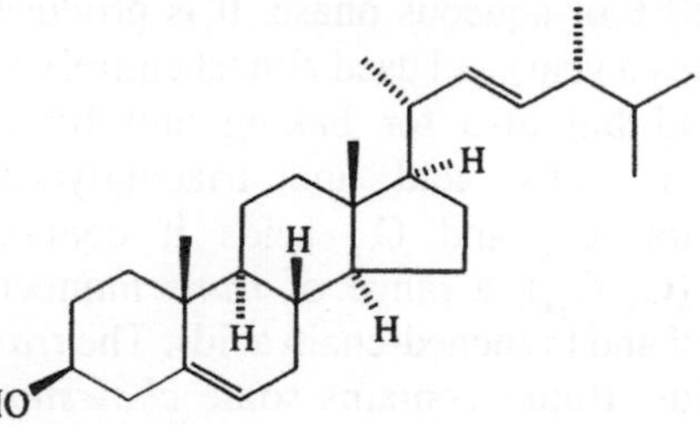

Brassicasterol
$C_{28}H_{46}O$
Mol. Wt.: 398.7

brassidic acid
the *trans (E)* form of *erucic acid* [22:1(n–9), mp 61.5°C].

brassylic acid
the C_{13} dibasic acid (tridecanedioic). See *dibasic acids*. This acid can be produced from *erucic acid* by *ozonolysis* and can be used commercially to produce a special nylon.

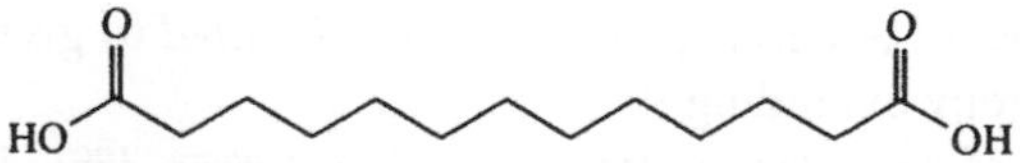

Brassylic acid
$C_{13}H_{24}O_4$
Mol. Wt.: 244.3

brown fat

brown adipose tissue which has the physiological function of heat generation in, for example, newborn humans and hibernating mammals. The brown colour, as opposed to normal white *adipose tissue,* comes from the cytochromes in the numerous mitochondria. It is known to be important in the generation of heat to maintain body temperature.

J. Himms-Hagen,Brown Adipose Tissue Thermogenesis and Obesity, Progress in Lipid Research, 1989, 28, 2, 67-115.

butolic acid

this acid, present in *shellac,* is 6-hydroxytetradecanoic acid.

butter

(i) a semi-solid material made from milk (mainly cow milk). Production is about 5.8 million tonnes a year on a fat basis. It is a water-in-oil emulsion containing 80–82% milk fat and 18–20% of aqueous phase. It is produced throughout the world (6–7 million tonnes a year) and used almost entirely for edible purposes, mainly as a spread but also for baking and frying. Butterfat is very complicated in its fatty acid and triacylglycerol composition. In addition to the usual C_{16} and C_{18} acids it contains short-chain and medium-chain acids (C_4–C_{14}), a range of *trans* monoene acids — mainly 18:1 — and oxygenated and branched-chain acids. The *trans* acids represent 4–8% of the total acids. Butter contains some *cholesterol* (0.2–0.4%). Spreads with lower levels of fat are also available. Butter that spreads directly from the refrigerator is made by removing some of its higher melting glycerol esters or by blending with a vegetable oil. See also *anhydrous milk fat.*

W. de Greyt et al., Food and non-food applications of milk fat, Lipid Technology, 1993, 5, pp.138-140.   Lipase-catalysed modification of milk fat, Lipid Technology, 1995, 7, pp.10-12. E. Flack, Lipid Technologies and Applications (ed. F.D. Gunstone and F.B. Padley) Marcel Dekker, New York (1997), pp.305-327. D. Hettinga, Bailey's Industrial Oil and Fat Products, (ed. Y.H. Hui) John Wiley (1996), Volume 3, pp.1-63.

(ii) name sometimes given to a solid fat, e.g. cocoa butter.

butter fat

the fat present in *butter* and free of aqueous phase. It is *fractionated* to give harder and softer fractions thereby extending its range of use.

W. de Greyt et al., Food and non-food applications of milk fat, Lipid Technology, 1993, 5, pp.138-140.   Lipase-catalysed modification of milk fat, Lipid Technology, 1995, 7, pp.10-12. E. Flack, Lipid Technologies and Applications (ed. F.D. Gunstone and F.B. Padley) Marcel Dekker, New York (1997), pp.305-327.

buttermilk
the aqueous phase formed in the production of butter from cream. Contains proteins and *membrane lipids* (phospholipids).

butter oil
produced from butter by melting and separation of the oily phase from the aqueous phase.

butylated hydroxy anisole
a synthetic *antioxidant* (4-methoxy-3-tert-butylphenol) which shows good solubility in fat and reasonable stability in fried and baked products. It is very effective with animal fats (which contain little or no natural antioxidant) but somewhat less so with vegetable fats (already protected in part by natural antioxidant). It shows marked synergism with *butylated hydroxy toluene* and *propyl gallate* and can be used in foods up to a level of 200 ppm (E number 320).
E.N. Frankel, Lipid Oxidation, The Oily Press, Dundee (1998).

butylated hydroxy toluene
a synthetic *antioxidant* (4-methyl-2,6-di-*tert*-butylphenol). It is less soluble than *butylated hydroxy anisole* and is not soluble in propylene glycol (a common solvent for antioxidants). It is synergistic with *butylated hydroxy anisole* but not with *propyl gallate* and can be used in foods up to a maximum level of 200 ppm (E number 321).

BHT
4-methyl-2,6-di-*tert*-butylphenol
$C_{15}H_{24}O$
Mol. Wt.: 220.4
E.N. Frankel, Lipid Oxidation, The Oily Press, Dundee (1998).

butyric acid
butanoic acid (4:0) is a short-chain saturated acid present in cow milk fat (~4% by weight equivalent to ~8.5 % on a molar basis and therefore present in about

25% of milk fat triacylglycerols), where it is present exclusively in the *sn*–3 position.

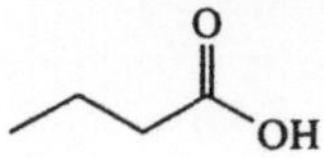

Butyric acid
butanoic acid
(4:0)
$C_4H_8O_2$
Mol. Wt.: 88.1

Butyrospermum parkii
a tree grown in Africa and Indonesia. The seed fat, called *shea butter* or karite, is rich in stearic (38–45%) and oleic (42–58%) acids. It contains a high level of *StOSt* among its glycerol esters and is used as a *cocoa butter equivalent*. Shea stearin prepared by fractionation is reported to contain 80% of StOSt.

C

calciferol
refers normally to vitamin D_2, ergocalciferol. See *vitamins* and *cholecalciferol.*

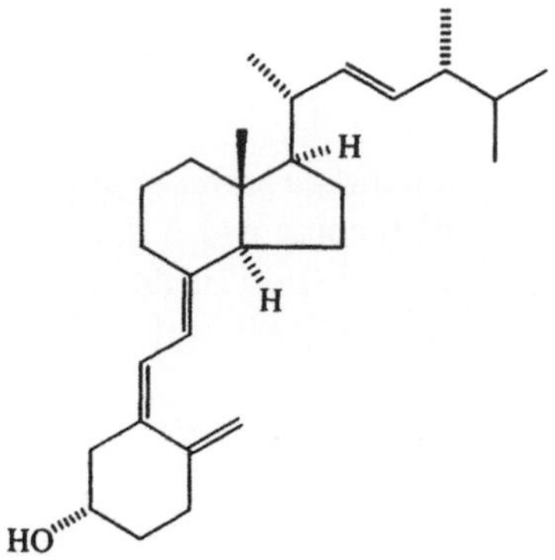

Ergocalciferol
$C_{28}H_{44}O$
Mol. Wt.: 396.6

calcium stearoyl lactate
see *sodium stearoyl lactate.* The calcium derivatives are less water dispersible than their sodium counterparts but more soluble in oils and fats.

calendic acid
see *conjugated unsaturation* and *Calendula officinalis.*

Calendula officinalis
also known as marigold. The oil from this seed oil (ca 19%) contains *calendic acid* (8t10t12c-octadecatrienoic acid 58%) and is a potential *drying oil.* Attempts are being made to develop this as a commercial crop, especially in the Netherlands.

Camelina sativa
also known as gold of pleasure or false flax. A new crop of potential interest

because it grows with low inputs of fertiliser and of pesticides. It contains
oleic acid (10–17%), linoleic acid (12–23%), α-linolenic acid
(31–41%), eicosenoic (12–18%), and docosenoic (2–4%) acids.

A. Hebard, Camelina sativa — a pleasurable experience or another false hope, Lipid Technology,
1998, 10, 81-83. E.C. Leonard, Camelina oil: α-linolenic source, INFORM, 1998, 9, 830-938.

campesterol
phytosterol common in plants.

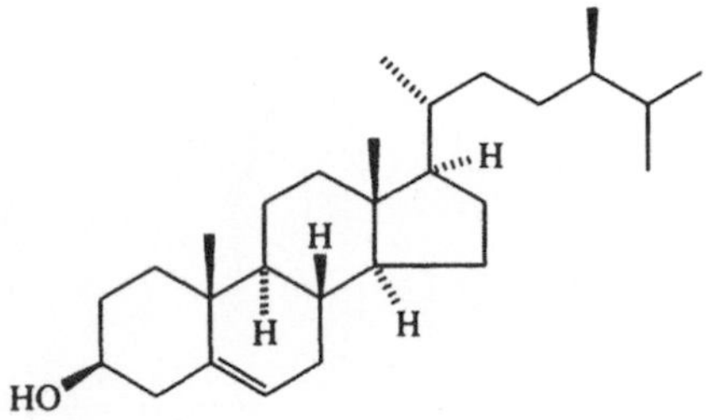

Campesterol
$C_{28}H_{48}O$
Mol. Wt.: 400.7

Candida species
see *lipases*.

canola oil
canola oil is obtained from interbred seeds of *Brassica napus* and *Brassica
campestris*. It is low in erucic acid and in glucosinolates and thus differs from
high-erucic rapeseed oil in its physical, chemical and nutritional properties.
Canola oil is defined by the Canadian Canola Council as oil from seed of the
genus *Brassica* with less than 1% of fatty acids in the oil as erucic acid. Unlike
the high-erucic variety, canola oil is widely used for food purposes.

N.A. Michael, B.E. Eskin, et al., Bailey's Industrial Oil and Fat Products, (ed. Y.H. Hui) John Wiley
(1996), Volume 2, pp.1-95.

capelin oil
a major type of *fish oil*. Typically capelin oil contains ~20% of saturated acids,
60% of monoene acids (16:1–22:1), and only modest levels of
eicosapentaenoic acid (9%) and docosahexaenoic acid (3%).

Caprenin ™
a semi-synthetic triacylglycerol with reduced caloric value. It contains equimolar proportions of *octanoic* (8:0), *decanoic* (10:0), and *behenic* (22:0, with some 20:0 and 24:0) acids corresponding to 22.6, 26.7, and 50.7% by weight. It consists mainly of triacylglycerols with 38, 40, and 42 carbon atoms. Its melting behaviour is similar to that of *cocoa butter* and it can be used in soft candy and in confectionery coatings for nuts, fruit, etc. The behenic acid is only partially absorbed (<20%) and caprenin has a calorific value of about 5 kcal/g as opposed to the usual value of 9 kcal/g. *J.W. Finley, et al., Lipid Technologies and Applications (ed. F.D. Gunstone and F.B. Padley) Marcel Dekker, New York (1997), pp.501-520*

capric acid
trivial name for *decanoic acid* (10:0, mp 31.6°C).

caproic acid
trivial name for *hexanoic acid* (6:0, mp –3.2°C).

caproleic acid
trivial name for 9-decenoic acid, a minor constituent of milk fat.

Caproleic acid
9-decenoic acid
$C_{10}H_{18}O_2$
Mol. Wt.: 170.2

caprylic acid
trivial name for *octanoic acid* (8:0, mp 16.5°C).

carboceric acid
trivial name for heptacosanoic acid (27:0, mp 82°C, 87.6°C).

cardiolipin
trivial name for *diphosphatidylglycerol*.

carnauba wax
a vegetable wax from *Copernica cerifera* containing about 30% of wax esters. These are mainly C_{46}–C_{54} esters based on C_{16}–C_{20} acids and C_{30}–C_{34} alcohols.

R.J. Hamilton, Waxes: Chemistry, Molecular Biology and Functions, (ed. R. J. Hamilton) Oily Press, Dundee (1995) pp.266-268.

carnitine

a β-hydroxy acid. Fatty acids (>C_{10}) are transported as acyl carnitines into the mitochondrion for β-oxidation.

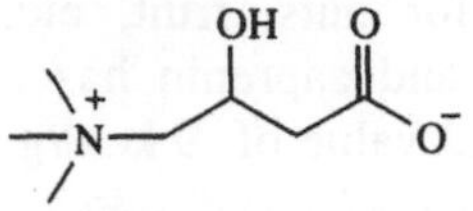

Carnitine
$C_7H_{15}NO_3$
Mol. Wt.: 161.2

carotene

see *carotenoids*. β-Carotene acts as a biological antioxidant, a free radical scavenger, a singlet oxygen quencher, a source of *vitamin A*, and as a source of colour in foods. It is also reported to have anti-cancer properties. There is some debate about which isomer is most active. It is largely removed from crude oil during refining but methods of retaining this valuable material have been developed to produce *red palm oil*. E number 160(a) covers α, β and γ-carotene.

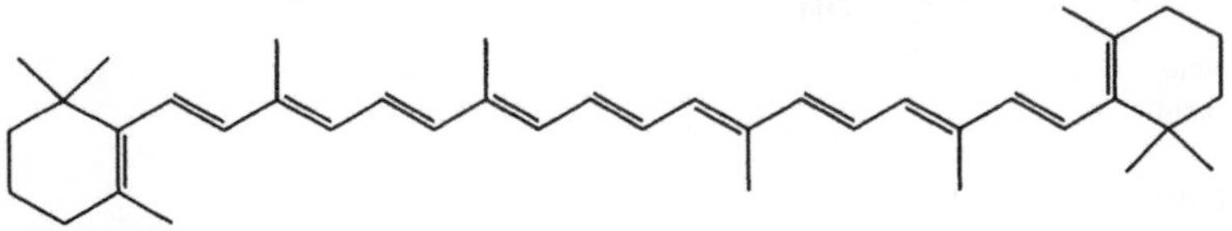

β-carotene
$C_{40}H_{56}$
Mol. Wt.: 536.9

carotenoids

a large group of isoprenoid structures with different numbers, positions and configurations of conjugated double bonds. The structure shown above is β-carotene, a precursor of vitamin A in animals. Carotenoids containing one or more hetero atoms (mainly oxygen) are known. Both α and β-carotene are reported to be anti-cancer agents. See *retinol* and *red palm oil*. A number of carotenoids are included under E numbers 160 and 161.

L.E. Schlipalius, Action mechanisms of carotenoids in the human body, Lipid Technology, 1997, 9,

39-43. F.C. Thyrion, Lipid Synthesis and Manufacture (ed. F.D. Gunstone) Sheffield Academic Press, Sheffield (1999) pp.268-287.

carthame (huile de)
French name for *safflower* seed oil.

Carthamus tinctorius
source of *safflower oil.*

castor oil
oil from *Ricinus communis* produced mainly in India, Brazil, and China. Production ~0.5 million tonnes a year. Castor oil differs from all other common oils in being rich (~90%) in a hydroxy acid, *ricinoleic* (12-hydroxy-*cis*-9-octadecenoic). Compared with other oils, castor oil is more viscous, less soluble in hexane, and more soluble in alcohol as a consequence of the presence of the hydroxy acid. Castor oil is a source of several important oleochemicals including *Turkey-red oil*, 12-hydroxystearic acid, *dehydrated castor oil*, heptanal, *10-undecenoic acid*, 2-octanol, and *sebacic acid*. See also *hydroxy acids.*

H-J. Caupin, Lipid Technologies and Applications (ed. F.D. Gunstone and F.B. Padley) Marcel Dekker, New York (1997), pp.787-795.

catalpic acid
see *conjugated unsaturation.*

cationic surfactants
surface-active compounds containing one or two long-chain alkyl groups attached to positively-charged nitrogen. They have many industrial uses.

A.D. James, Lipid Technologies and Applications (ed. F.D. Gunstone and F.B. Padley) Marcel Dekker, New York (1997), pp.609-631.

Cetyl trimethyl ammonium bromide
$C_{19}H_{42}BrN$
Mol. Wt.: 364.4

CBA
see *cocoa butter alternatives.*

CBE
see *cocoa butter equivalents*.

CBI
see *cocoa butter improvers*.

CBR
see *cocoa butter replacers*.

CBS
see *cocoa butter substitutes*.

cephalin
old term for phosphatidylethanolamine.

Ceramide-3 ™
the N-stearoyl derivative of *phytosphingosine*. It has cosmetic applications.

ceramide mono-, di- and polyhexosides
see *glycosyl ceramides*.

ceramides
trivial name for the lipid class N-acylsphingosines. The ceramides are part of the human skin protective barrier and are the building block of the complex *sphingolipids*. They are widely used in cosmetics.

Typical ceramide

cerebronic acid
2-hydroxytetracosanoic acid (mp 100°C for the *R,S*- form) is present in *cerebrosides* in the *R*- form. It has been isolated from the *glycosphingolipids* of wheat, corn, other plant species, and some microorganisms. It was formerly known as phrenosic and phrenosinic acid.

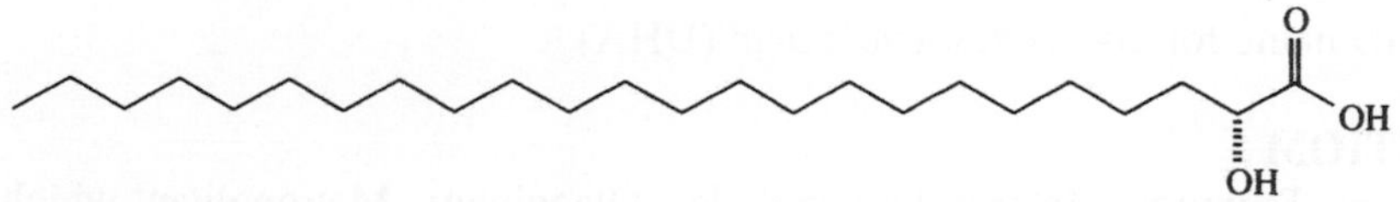

Cerebronic acid
2R-hydroxytetracosanoic acid
$C_{24}H_{48}O_3$
Mol. Wt.: 384.6

cerebrosides
trivial name for the lipid class monoglycosyl ceramides. Present in the myelin sheath of nerve and brain as monogalactosyl and monoglucosyl ceramides. See also *glycosyl ceramides*.

K.-H. Jung and R.R. Schmidt, Lipid Synthesis and Manufacture (ed. F.D. Gunstone) Sheffield Academic Press, Sheffield (1999) pp.208-249.

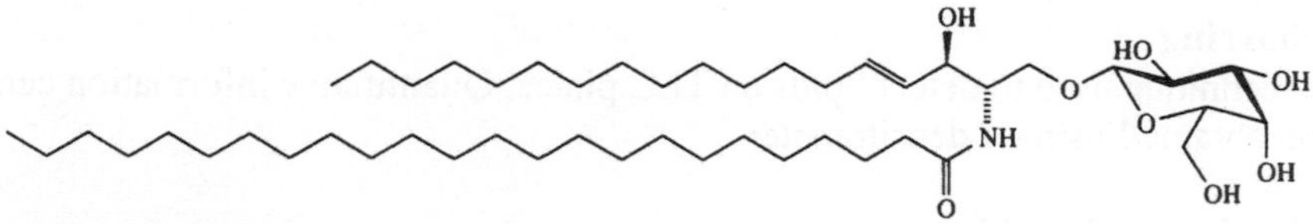

Typical cerebroside

ceromelissic acid
trivial name for tritriacontanoic acid (33:0).

ceroplastic acid
trivial name for pentatriacontanoic acid (35:0).

cerotic acid
trivial name for hexacosanoic acid (26:0, mp 87.8°C).

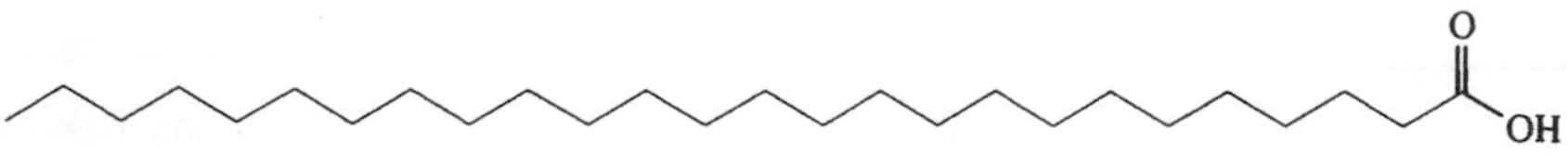

Cerotic acid
Hexacosanoic acid
(26:0)
$C_{26}H_{52}O_2$
Mol. Wt.: 396.7

cervonic acid
trivial name for *docosahexaenoic acid* (DHA).

CETIOM
Centre Technique Interprofessional des Oleagineux Metropolitan which supports development of oilseeds in France.

cetoleic acid
trivial name for *cis*-11-docosenoic acid, (22:1 *n*-11, mp 33–33.7°C). Present in many fish oils.

chain elongation
biosynthetic conversion of a fatty acid as its co-enzyme A derivative to the bishomologue by addition of a C_2 unit by means of an *elongase*.
J.L. Harwood, Lipid Synthesis and Manufacture (ed. F.D. Gunstone) Sheffield Academic Press, Sheffield (1999) pp.422-466.

charring
a technique used to detect lipids on TLC plates. Quantitative information can be obtained using a densitometer.

chaulmoogric acid
trivial name for 13-(2-cyclopentenyl)tridecanoic acid (*S*-form, mp 67–68°C). See also *cyclopentenyl acids*

Chaulmoogric acid
13-(2-cyclopenten-1*S*-yl)tridecanoic acid
$C_{18}H_{32}O_2$
Mol. Wt.: 280.4

chelator
substances such as *citric acid* and *EDTA* that combine with metal ions making these unable to act as pro-oxidants.
E.N. Frankel, Lipid Oxidation, The Oily Press, Dundee (1998).

Chemistry and Physics of Lipids
journal published by Elsevier Scientific Publishers Ireland Ltd since 1966.

cherry oil

the oil from cherry kernels (*Prunus* spp.) containing significant levels of oleic (ca 35%) and linoleic (ca 45%) acid and also *eleostearic acid* (ca 10%).

chimyl alcohol

trivial name for the glycerol ether 1-*O*-hexadecyl-*sn*-glycerol (mp 64–65°C) present along with other glycerol ethers (*batyl alcohol, selachyl alcohol*) in some marine liver oils as a diacylated derivative.

Chimyl alcohol
S-1-O-hexadecyl-*sn*-glycerol
$C_{19}H_{40}O_3$
Mol. Wt.: 316.5

China wood oil

see *tung oil*.

chinese vegetable tallow

from the tree *Sapium sebiferum (Stillingia sebifera)*. The fruit furnishes fatty material from the outer seed coating (chinese vegetable tallow) and also from the seeds (*stillingia oil*). The first of these is highly saturated (palmitic acid 62% and oleic acid 27%) whilst the second has a very unusual composition including an *estolide*.

chloroplasts

photosynthetic organelles in plant cells where fatty acids are metabolised. In many respects they are equivalent to the *mitochondria* in animal cells.

3-chloropropanediol

a compound formed from glycerol by reaction with hydrochloric acid. It is formed during protein hydrolysis with hydrochloric acid from lipid present in the protein sample.

chocolate

product made from cocoa beans that typically contains 10–15% cocoa, 30–50% sugar and 27–35% fat. Milk powder is added to get milk chocolate and blended cocoa beans to get dark chocolate. The fat (*cocoa butter* or *cocoa butter alternative*) is an important component since its characteristic melting behaviour contributes to the acceptable mouthfeel of chocolate. It is also the most expensive component in the chocolate.

F.B Padley, Lipid Technologies and Applications (ed. F.D. Gunstone and F.B. Padley) Marcel Dekker, New York (1997), pp.391-432.

cholecalciferol

vitamin D_3, a compound with anti-rachitic properties. It is present in some foods (high content in cod liver oil) and is produced under the influence of sunlight by transformation of 7-dehydrocholesterol in skin surface lipids.

Cholecalciferol
(3β,5Z,7E)-9,10-Secocholesta-5,7,10-(19)-trien-3-ol
$C_{27}H_{44}O$
Mol. Wt.: 384.6

cholesterol

the most common animal sterol present in free or esterified form. It is important in *membranes* and *lipoproteins* and serves as a precursor of hormones, *bile acids* etc. Cholesterol concentration in blood plasma is around 6 mM (equivalent to about 230 mg/100 ml). The human body contains about 100 g of cholesterol. Average daily consumption of cholesterol in the UK is ~300 mg. The normal level in vegetable oils (maximum 10–20 ppm) is in marked contrast to the higher values observed in animal fats such as *lard* (0.37–0.42%), mutton *tallow* (0.23–0.31%), *beef fat* (0.08–0.14%), and *butter* fat (0.2–0.4%). Hen eggs contain ~300 mg of cholesterol per egg (~5% of total lipid).

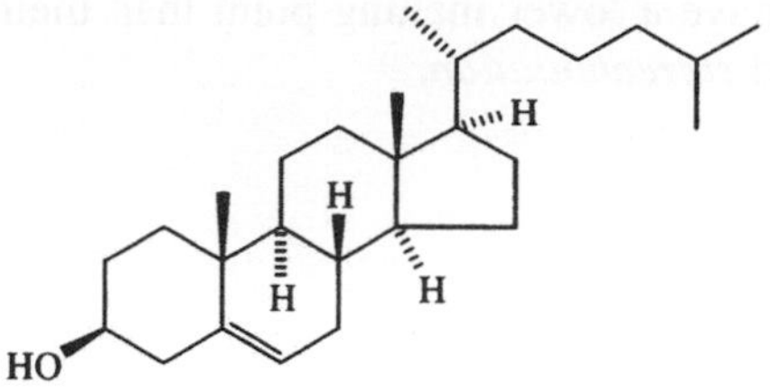

Cholesterol
Cholest-5-en-3β-ol
$C_{27}H_{46}O$
Mol. Wt.: 386.7

cholesterol esters
cholesterol esterified to fatty acids. Present in animal tissues and *lipoproteins*.

cholic acid
see *bile acids, bile salts*.

chromatography
separation technique based on two immiscible phases, the stationary phase (e.g. a column packing or a thin-layer) and the mobile phase, normally a liquid or a gas. The mixture to be separated is carried by the mobile phase through the stationary phase. Because of different affinities (adsorption, partition) for the stationary phase the components of the mixture are delayed to different degrees compared to the velocity of the mobile phase. See also *gas chromatography, liquid chromatography and thin-layer chromatography*.

Chromobacterium
see *lipases*.

chrysobalanic acid
the trivial name for 4-oxo-9c11t13t15c-18:4. This acid is present in *Chrysobanus icaco* seed oil.

chylomicrons
see *lipoproteins*.

cis double bond
compounds containing a carbon–carbon double bond can exist in two

stereoisomeric forms (*cis* or *trans*). Natural unsaturated fatty acids generally have the *cis* (*Z*) configuration. They have a lower melting point than their *trans* isomers. See also *trans acids* and *stereomutation*.

Cis double bond

Citrem ™

citric acid esters of *monoacylglycerols* formed by reaction of citric acid with mono-/di-acylglycerol mixtures or with distilled monoacylglycerol. The acidic product is partially neutralised with sodium hydroxide. A typical product contains 60–90% monoacylglycerol and ~20% citric acid (minimum 12%). Citrems are used in frying *margarines* as anti-spattering agents and in meat emulsions to inhibit fat separation. E number 472 (c) in Europe and US/FDA/CFR 172832.

N. Krog, Lipid Technologies and Applications (ed. F.D. Gunstone and F.B. Padley) Marcel Dekker, New York (1997), pp.521-534.

citric acid

a monohydroxy tricarboxylic acid used in the refining of oils and fats to remove trace metals by chelation.

K.S. Law et al., Citric acid in the processing of oils and fats, PORIM Technology, No. 11, 1984.

Citric acid
2-Hydroxy-1,2,3-propanetricarboxylic acid
$C_6H_8O_7$
Mol. Wt.: 192.1

CLA

see *conjugated linoleic acid*.

clofibrate

drug, used because of its lowering effect on blood lipids (*triacylglycerols* and *cholesterol*).

Clofibrate
Ethyl 2-(4-chlorophenoxy)-2-methylpropanoate
$C_{12}H_{15}ClO_3$
Mol. Wt.: 242.7

clupadonic acid
the name given to a 22:5 acid. It was reported to be a 4,8,12,15,19-pentaenoic acid but it is now known to be the 7,10,13,16,19 (n–3) isomer.

CMC
see *critical micellar concentration.*

[13]C nuclear magnetic resonance spectroscopy
[13]C *nuclear magnetic resonance* (NMR) spectra are complex but contain useful structural information and can distinguish between acyl groups in the α and β positions of triacylglycerols. *Cis* and *trans* isomers are easily distinguished by their allylic and olefinic signals. Less common functional groups (epoxy, hydroxy, branched methyl etc) can also be recognised.

F.D. Gunstone, Advances in Lipid Methodology — Two (ed. W.W. Christie), The Oily Press, Dundee, 1993, pp.1-68. M.S.F. Lie Ken Jie and J. Mustafa, High-resolution nuclear magnetic resonance spectroscopy — applications to fatty acids and triacylglycerols, Lipids, 1997, 32, 1019.

CNRS
Centre National de la Recherche Scientifique. A French agency devoted to fundamental research.

cocoa
the powder manufactured from cocoa beans from which the shell and some of the fat (*cocoa butter*) have been removed.

cocoa butter
the cocoa bean (*Theobroma cacao*) is the source of two important ingredients of chocolate: cocoa powder and a solid fat called cocoa butter. The usefulness of cocoa butter for this purpose is related to its fatty acid and triacylglycerol composition. The major triacylglycerols are symmetrical disaturated oleic

glycerol esters of the type SOS and include POP (18–23%), POSt (36–41%), and StOSt (23–31%). Cocoa butter commands a good price and cheaper alternatives have been developed (see following entries and also *palm mid fraction, Illipe, sal*. The annual production of cocoa beans is about 2.7 million tonnes with 45–48% of cocoa butter.

V.K.S. Shukla. Cocoa butter properties and quality, Lipid Technology, 1995, 7, 54-57. F.B Padley, Lipid Technologies and Applications (ed. F.D. Gunstone and F.B. Padley) Marcel Dekker, New York (1997), pp.391-432. P.J. Lawler and P.S. Dimick, Food Lipids: Chemistry, Nutrition, and Biotechnology, (ed. C.C. Akoh and D.B. Min) Marcel Dekker, New York (1998) pp.229-250.

cocoa butter alternatives
general term covering cocoa butter equivalents, cocoa butter improvers, cocoa butter replacers, and cocoa butter substitutes. See also *cocoa butter* and *confectionery fats*.

cocoa butter equivalents
fats with the same type of triacylglycerol composition as cocoa butter. See also *cocoa butter alternatives*, *cocoa butter* and *confectionery fats*.

cocoa butter improvers
fats with the same type of triacylglycerol composition as cocoa butter. See also *cocoa butter alternatives*, *cocoa butter* and *confectionery fats*.

cocoa butter replacers
products that can partially replace cocoa butter in confectionery formulations. Normally fractionated fats, e.g. *palm oil* or hydrogenated *soybean oil* fractions. See also *cocoa butter alternatives*, *cocoa butter* and *confectionery fats*.

cocoa butter substitutes
Similar to *cocoa butter replacers* but generally based on coconut and palm kernel oils. See also *cocoa butter alternatives*, *cocoa butter* and *confectionery fat*.

coconut oil
a major *lauric oil* obtained from *copra* which is a product of the coconut palm (*Cocus nucifera*). Coconut oil (~3.1 million tonnes per annum) comes mainly from Indonesia and the Philippines. It is particularly rich in *lauric acid* (~47%) and *myristic acid* (~18%). Also present are 8:0 (8%) and 10:0 (7%) which are easily separated by distillation and used to make *medium-chain triglycerides, Caprenin™*, Miglyol™ etc. The oil finds extensive use in the

food industry and also — usually after conversion to the alcohol *(dodecanol etc)* — in the detergent, cosmetic, and pharmaceutical industries. The only other commercially available lauric oil is *palm kernel oil* but see also *laurate-canola* and *cuphea* species.

E.C. Canapi. Y.T.V. Augustin et al., Bailey's Industrial Oil and Fat Products, (ed. Y.H. Hui) John Wiley & Sons, New York (1996), Volume 2, pp.97-124.

Cocus nucifera
see *coconut oil.*

Codex Alimentarius

an organisation under the Food and Agriculture Organisation and the World Health Organisation whose fats division publishes standard specifications for all major edible oils and fats.

cod liver oil
an important fish oil which serves as a valuable source of vitamins A and D. It contains saturated (~20%), monoenoic (16:1–22:1, 55–60%), and eicosapentaenoic and docosahexaenoic acids (each ~ 7%). See *fish oils.*

coffee whitener
a substitute for dairy cream in coffee. It generally contains a hardened lauric oil.

colneleic acid
an ether C_{18} acid produced during enzymic oxidation of potato lipids. It is derived from *linoleic acid* by rearrangement of its 9-*hydroperoxide*. A similar product (*colnelenic acid*) is based on linolenic acid and other acids of this type have also been recognised.

T. Galliard et al., Novel divinyl ether fatty acids in extracts of Solanum tuberosum, *Chem. Phys. Lipids, 1973, 11, 173-180. L. Crombie et al., An isotopic study of the enzymic conversion of linoleic acid into colneleic acid and carbon chain fracture: the origin of shorter chain aldehydes. J. Chem. Soc. (Perkin I) 1991, 567-575. A. Grechkin, Prog. Recent developments in the biochemistry of the plant lipoxygenase pathway, Progress in Lipid Research, 1998, 37, 317-352.*

Colneleic acid
9-(1*E*,3*Z*-Nonadienyloxy)-8*E*-nonenoic acid
$C_{18}H_{30}O_3$
Mol. Wt.: 294.4

colnelenic acid

an ether C_{18} acid produced during enzymic oxidation of potato lipids. It is derived from *linolenic acid* by rearrangement of its *9-hydroperoxide*. A related product (*colneleic acid*) is based on linoleic acid and other acids of this type have also been recognised.

T. Galliard et al., Novel divinyl ether fatty acids in extracts of Solanum tuberosum, Chem. Phys. Lipids, 1973, 11, 173-180. L. Crombie et al., An isotopic study of the enzymic conversion of linoleic acid into colneleic acid and carbon chain fracture: the origin of shorter chain aldehydes, J. Chem. Soc. (Perkin I) 1991, 567-575. A. Grechkin, Prog. Recent developments in the biochemistry of the plant lipoxygenase pathway, Progress in Lipid Research, 1998, 37, 317-352.

Colnelenic acid
9-(1E,3Z,6Z-Nonadienyloxy)-8E-nonenoic acid
$C_{18}H_{28}O_3$
Mol. Wt.: 292.4

columbinic acid

see *octadecatrienoic acid.*

colza oil

an old name for (high-erucic) *rapeseed oil*, and which is still used in the French language.

complex lipids

lipids which on hydrolysis yield at least three building blocks, e.g. fatty acids, phosphoric acid, amino alcohols, sugars and glycerol from phospholipids or glycolipids. See also *simple lipids.*

confectionery fats

special fats for confectionery applications, in combination with or as a replacement for *cocoa butter*, generally termed *cocoa butter alternatives.* Cocoa butter equivalents have the same major triacylglycerols as cocoa butter (e.g. fractionated *palm oil, shea fat*). Cocoa butter substitutes or cocoa butter replacers have similar physical properties but different triacylglycerol composition (e.g. fractionated and hydrogenated *palm kernel oil, soybean oil, rapeseed oil*). The use of these materials is controlled by regulations but not uniformly in all countries.

F.B Padley, Lipid Technologies and Applications (ed. F.D. Gunstone and F.B. Padley) Marcel Dekker, New York (1997), pp.391-432.

conjugated linoleic acid (CLA)
the name given to a mixture of C_{18} diene acids with conjugated unsaturation. Rumen hydrogenation gives a mixture with the $9c11t$ isomer (*rumenic acid*) predominating. It is present at low levels in products from the meat or milk of ruminants. Interest in this mixture has developed following claims that CLA has many beneficial effects including inhibition of cancer development and promotion of muscle development at the expense of fat. The active compound is thought to be the $9c11t$ isomer (rumenic acid) but this has not been proved. This isomer can be synthesised from methyl ricinoleate. Alkaline isomerisation of linoleic acid gives a mixture of 9,11 and 10,12 dienes with many minor components and this is being produced commercially for use as conjugated linoleic acid.

S. Banni and J.-C. Martin, Trans Fatty Acids in Human Nutrition (eds J.L. Sebedio and W.W. Christie) The Oily Press, Dundee (1998) pp.261-302. M.P. Jurawecz (eds) Advances in Conjugated Linoleic Acid Research, Vol.1, AOCS Press, Champaign, Illinois (1999).

conjugated unsaturated acids
when two or more unsaturated centres are immediately adjacent to each other they are said to be conjugated. Natural acids with conjugated unsaturation are mainly C_{18} trienes or tetraenes such as jacaric ($8c10t12c$, mp 43.5–44°C) in *Jacaranda mimosifolis* seed oil, calendic ($8t10t12c$, mp 40–40.5°C) in *Calendula officinalis* seed oil, catalpic ($9c11t13c$, mp 31.5–32°C) in *Catalpa avata* seed oil, α-eleostearic ($9c11t13t$, mp 48–49°C) in *Aleurites* oils, punicic ($9c11t13c$, m.p, 44–45°C) in *Punica granatum* seed oil, β-eleostearic ($9t11t13t$, mp 72°C), α-parinaric ($9c11t13t15c$, mp 72–74°C) in *Impatiens balsamina* seed oil and β-parinaric acid ($9t11t13t15t$, mp 95–96°C). Other acids with conjugated unsaturation also contain hydroxy or oxo groups.

convolvulinic acid
this is probably 11-hydroxytetradecanoic acid but the name has been used also for 3,12-dihydroxytetradecanoic acid. It is present in *Ipomea* oils.

cooling curve
a standard in the physical characterization of fats. It is a plot of time against temperature of a sample of oil in a cooling bath and indicates phase transitions. See *polymorphism*.

copra
see coconut oil.

coriander oil

coriander *(Coriandrum sativum)* contains about 80% of *petroselinic acid* in its seed oil. Attempts are being made to develop better sources of this acid through improved cultivation of coriander or by transfer of appropriate genes from coriander to *rape*.

coriolic acid

13-hydroxy $9c11t$-octadecadienoic acid. It occurs in rare seed oils in the $R(-)$, $S(+)$, and $RS(\pm)$ forms. It is readily dehydrated to conjugated $\Delta9,11,13$ and $\Delta8,10,12$ acids and is an isomer of *dimorphecolic acid*. It also occurs as a lactone *(coriolide)*.

(-)-Coriolic acid
13R-Hydroxy-9Z,11E-octadecadienoic acid
$C_{18}H_{32}O_3$
Mol. Wt.: 296.4

coriolide

the lactone of *coriolic acid* which co-exists with the hydroxy acid in *Monnina emarginata* seed oil.

corn oil

a major vegetable oil (~2.0 million tonnes per annum) from corn or maize *(Zea mays)* obtained by wet milling, particularly in the USA. The major acids are palmitic (9–17%), oleic (20–42%), and linoleic (39–63%) and the major triacylglycerols are typically LLL (15), LLO (21), LLS (17), LOO (14), LOS (17), LSS (5), OOO (6), and OOS (4%). Despite its high level of unsaturation the oil has good oxidative stability. A high-oleic variety is being developed.

L.R. Strecker, M.A. Bieber, et al., Bailey's Industrial Oil and Fat Products, (ed. Y. H. Hui) John Wiley & Sons, New York (1996), Volume 2, pp.125-158.

coronaric acid

a rare natural *epoxy acid* isomeric with the more common *vernolic acid*. The (9R10S)-epoxy-12-*cis*-octadecenoic acid is present in *Chrysanthemum coronarium* and some other seed oils. It inhibits the growth of rice blast fungus. It is a *leukotoxin* produced in plants from *linoleic acid*.

Coronaric acid
9*R*,10*S*-Epoxy-12*Z*-octadecenoic acid
$C_{18}H_{32}O_3$
Mol. Wt.: 296.4

coton (huile de)
French name for *cottonseed* oil.

cottonseed oil
a major vegetable oil (4.0 million tonnes per annum) obtained as a by-product in the production of cotton and grown mainly in China, USA, the former Soviet Union, India, and Pakistan. It ranks fifth among vegetable oils. Cottonseed oil is rich in palmitic acid (22–26%), oleic acid (15–20%) and linoleic acid (49–58%) but also contains some C_{20}–C_{24} acids (about 1%) and two cyclopropene acids (*sterculic* and *malvalic*). The latter are identified by the *Halphen test*. They are removed during refining. One sample of cottonseed oil contained the following triacylglycerols: PLL (26), LLL (16), POL (14), OLL (13), PLP (9) and seven other components (22%). Among the minor constituents of the oil is the yellow pigment *gossypol*.

L.A. Jones and C.C. King, *Bailey's Industrial Oil and Fat Products*, (ed. Y.H. Hui) John Wiley & Sons, New York (1996), Vol.2, pp.159-240.

couepic acid, couepinic acid
other names, now unused, for *licanic acid*.

crambe oil
oil from *Crambe abyssinica* and *C. hispanica*, characterised by a high level of *erucic acid* (55–60%) and grown almost exclusively in North Dakota (USA). See also high-erucic *rapeseed oil* (HEAR). Genetically modified erucic-rich oils are being developed in response to the demand for erucic acid and its derivatives (*erucamide, behenic acid, brassylic acid, behenyl alcohol, and erucyl alcohol*).

E.C. Leonard, *Sources and commercial applications of high-erucic vegetable oils, Lipid Technology, 1994, 6, 79-83.*

crepenynic acid
an acetylenic analogue of linoleic acid, 9c12a-18:2, present in *Crepis* and *Afzelia* oils. It is a key intermediate in the biosynthesis of the large group of natural short-chain and medium-chain acetylenic compounds.

Crepenynic acid
9Z-Octadecen-12-ynoic acid
$C_{18}H_{30}O_2$
Mol. Wt.: 278.4

critical pair
term formerly used in reversed-phase partition chromatography of lipids indicating the co-elution of components with the same *partition number*, e.g. 18:1/16:0 methyl esters, triolein/tripalmitin etc. These are now generally resolvable under suitable conditions.

Crypthecodinium cohnii
an alga producing lipid rich in *docosahexaenoic acid* (see *Dhasco™*).

crystallization
see *fractionation*.

crystal structure of triacylglycerols
triacylglycerols display *polymorphism*, i.e. they can exist in more than one crystalline form. These differ in melting point, in physical stability, and in the arrangement of molecules within the crystal. When a liquid triacylglycerol is cooled quickly it solidifies in an α crystalline form. Under appropriate temperature conditions this changes first to a β′ form and then to a β form. The β form has the highest melting point of the three forms, is the most stable form, and is the form obtained by crystallisation from a solvent. The three crystalline forms can be distinguished by X-ray crystallography or by their infrared spectra. The long spacings of the crystals indicate that these generally have double chain length (DCL) i.e. two acyl chains per unit cell but some molecules (e.g. POP) exist in a triple chain length form (TCL) i.e. three acyl chains per unit cell because this allows more efficient packing.
β′ Crystals are relatively small and can incorporate a large amount of liquid. *Margarine* and *shortenings* containing crystals in this form have a glossy surface and a smooth texture. β Crystals, though initially smooth, grow into

needle-like agglomerates less able to incorporate liquid and producing a grainy texture. β' Crystals are more likely to be the stable form in triacylglycerols of differing chain length such as *cottonseed oil* and *palm oil* where significant levels of palmitic acid accompany the C_{18} acids. Oils with low levels of palmitic acid and high levels of C_{18} acids (canola, soybean, sunflower) tend to exist as β crystals (after partial hydrogenation) though there are ways of inhibiting the β' to β change.

K. Larsson, Lipids — Molecular Organisation, Physical Functions and Technical Applications, The Oily Press, Dundee (1994). P.J. Lawler and P.S. Dimick, Food Lipids: Chemistry, Nutrition, and Biotechnology, (ed. C.C. Akoh and D.B. Min) Marcel Dekker, New York (1998) pp.229-250.

CSL

abbrevation for *calcium stearoyl lactate.*

cubic phase

lyotropic (lipid-water) liquid crystalline phase, consisting of a three dimensional network of lipid bilayer walls separating two water channel systems.

K. Larsson, Lipids — Molecular Organisation, Physical Functions and Technical Applications, The Oily Press, Dundee (1994).

cuphea

wild plant species characterised by the presence of glycerol esters rich in medium-chain acids (C_8-C_{14}). Attempts are being made to domesticate some of the cuphea species as additional lauric oils and as rich sources of other medium-chain acids (e.g. *C. viscosissima* and *C. lanceolata* with 75–85% of decanoic acid).

cutin

a polymer of hydroxy acids which serves as an outer envelope for plants. The acids are mainly C_{16} (including palmitic, 16-hydroxypalmitic, and 10,16-dihydroxypalmitic) and C_{18} (including oleic, 18-hydroxyoleic, 18-hydroxy-9,10-epoxyoctadecanoic, 9,10,18-trihydroxyoctadecanoic, and phloinolic).

M. Riederer and L. Schreiber, Waxes: Chemistry, Molecular Biology and Functions, (ed. R.J. Hamilton), The Oily Press, Dundee (1995) p.131.

cyclic acids

though not common, such acids occur naturally (see *cyclopropene* and *cyclopentene* acids). They are also formed during processes, such as frying, in which unsaturated acids are exposed to high temperature. One double bond is lost in the cyclisation process. For example, monocyclic monoenes resulting from linoleate contain cyclopentane or cyclohexane units. The residual double bond may be exocyclic or endocyclic.

G. Dobson, *Recent Developments in the Synthesis of Fatty Acid Derivatives (ed. G. Knothe and J.H.P. Derksen) AOCS Press, Champaign (1999), pp.196-212.*

cyclohexyl acids

fatty acids with a terminal cyclohexane group. Such acids have been identified in certain bacteria and in sheep fat and butter.

Typical cyclohexyl acid

cyclopentenyl acids

fatty acids with a terminal cyclopentenyl group. Seed fats of the Flacourtiaceae (e.g. chaulmoogra oil used in folk medicine for the treatment of leprosy) are unique in producing cyclopentene acids with one or two double bonds. The monoene acids include aleprolic ($n = 0$), alepramic ($n = 2$), aleprestic ($n = 4$), aleprylic ($n = 6$), alepric ($n = 8$), hydnocarpic ($n = 10$, mp 58–59°C), chaulmoogric ($n = 12$, mp 67–68°C) and hormelic ($n = 14$). The diene members include manaoic ($x = y = 4$), gorlic ($x = 4; y = 6$, mp 6°C) and oncobic ($x = y = 6$).

Typical cyclopentyl acid

cyclopropane acids

fatty acids with a cyclopropane unit in the chain. They occur in membrane phospholipids of certain bacteria (e.g. *lactobacillic acid*).

Cyclopropane acid

cyclopropene acids

the best known cyclopropene acids are *malvalic* (C_{18}) and *sterculic* (C_{19}) which are present at high levels in sterculia oils and at lower levels in *kapok seed oil* (~12 %) and in *cottonseed oil* (~1%). These highly reactive acids are destroyed during refining and during hydrogenation. They have attracted interest because they inhibit the biodesaturation of stearic to oleic acid. Hydroxy (*2-hydroxysterculic*) and acetylenic (*sterculyinic*) analogues also occur.

Cyclopropene acid

cytosides
trivial name for *diglycosylceramides.*

D

DAG
see *diacylglycerols*.

Datem ™
diacetyltartaric acid esters of mono-/di-glycerides and of distilled monoglycerides made by reacting glycerol esters with diacetyltartaric anhydride. These compounds are anionic, hydrophilic, water-dispersible emulsifiers used as dough-strengtheners in yeast-raised bakery products such as bread. E number 472 (e) in Europe.
N. Krog, Lipid Technologies and Applications (ed. F.D. Gunstone and F.B. Padley) Marcel Dekker, New York (1997), pp.521-534.

daturic acid
an old trivial name for *heptadecanoic* (margaric) acid.

DCL
double chain length. See *crystal structure*.

DCO
see *dehydrated castor oil*.

decanoic acid
the C_{10} acid (caproic) which occurs in *coconut oil, palm kernel oil* and some *Cuphea oils*. It is a minor component of milk fat and an important acid in *medium chain triglycerides, Caprenin™* and Miglyol™·.

degumming
an early step in the refining oils and fats. Addition of ~2% of water (sometimes containing phosphoric acid) at 70–80°C to the crude oil results in the separation of most of the phospholipids accompanied by trace metals and pigments. The insoluble material that is removed is mainly a mixture of *phospholipids* and *triacylglycerols* and is known as *lecithin*. It is obtained mainly from soybean refining and is a valuable source of phospholipids used to obtain a range of higher grade products.
D.A. Allen, Lipid Technologies and Applications (ed. F.D. Gunstone and F.B. Padley) Marcel Dekker, New York (1997), pp.137-167.

dehydrated castor oil

dehydration of castor oil gives a product low in *ricinoleic acid* and enriched in conjugated and non-conjugated C_{18} diene acids (mainly stereoisomers of the 9,11- and 9,12-dienes). It is used as a *drying oil*.

demospongic acids

a group of C_{24}–C_{34} acids present in many *sponge* species. These acids are characterised by the presence of 5c9c unsaturation often accompanied by other functionality and occur mainly in the phospholipids. Two demospongic acids may occur in the same phospholipid molecule. Additional features sometimes present include a branched methyl group, cyclopropane unit, α-hydroxy, acetoxy or methoxy group, and 6-bromo group. Sometimes 5,9 unsaturation is replaced by 5,11 unsaturation or the 5,9 system can be elongated to 7,11 etc. See also *non-methylene-interrupted polyene acids*.

C. Djerassi et al., Sponge phospholipids, Acc. Chem. Res. 1991, 24, 69-75.

de novo fatty acid synthesis

the biosynthetic pathway by which acetate (as its *ACP* ester) is converted to medium-chain and long-chain saturated acids, especially palmitic, by a fatty acid synthetase. In plant systems this occurs in the plastid.

K.L. Parkin, Food Lipids: Chemistry, Nutrition, and Biotechnology, (ed. C.C. Akoh and D.B. Min) Marcel Dekker, New York (1998 pp.729-778. J.L. Harwood, Lipid Synthesis and Manufacture (ed F.D. Gunstone) Sheffield Academic Press, Sheffield (1999) pp.422-466.

densipolic acid

a rare hydroxy acid present in *Lesquerella densipila* seed oil. It is an unsaturated analogue of the more common *ricinoleic acid*.

Densipolic acid
12R-Hydroxy-9Z,15Z-octadecadienoic acid
$C_{18}H_{32}O_3$
Mol. Wt.: 296.4

deodorization

a treatment of oils and fats at a high temperature (200–250°C) and low pressure (0.1–1 mm Hg). It is an important step in the refining of oils and fats resulting in the removal of volatile and odorous compounds including *free fatty acids, monoacylglycerols* and oxidation products. At the high temperatures involved in this process there is a danger of *stereomutation*, especially of linolenic acid. Refined oils containing this acid generally contain significant levels of *trans* isomers.

P. Sjoberg, Deodorization technology, Lipid Technology, 1991, 3, 52-57. D.A. Allen, Lipid Technologies and Applications (ed. F.D. Gunstone and F.B. Padley) Marcel Dekker, New York (1997), pp.137-167.

deodorizer distillate

a condensation product recovered during the *deodorization* process in which oil is sparged with steam at high temperatures. The deodorizer distillate is a valuable source of *tocopherols* (vitamin E) and of *phytosterols* (2000-4000 ppm).

J.P. Clark, Tocopherols and sterols from soybeans, Lipid Technology, 1996, 8, 111-114. T. Netscher, Lipid Synthesis and Manufacture (ed. F.D. Gunstone) Sheffield Academic Press, Sheffield (1999) pp.250-267.

de-oiling

a process for removing most of the triacylglycerols from crude *lecithin* by heating with acetone. The level of phospholipid is raised from 45–50% to 95–98% by this process.

M. Schneider, Industrial production of phospholipids — lecithin processing, Lipid Technology, 1997, 9, pp.109-116. M. Schneider, Lipid Technologies and Applications (ed. F.D. Gunstone and F.B. Padley) Marcel Dekker, New York (1997), p.51.

deoxycholic acid

see *bile acids, bile salts*.

depot fat

general term for the *triacylglycerols* stored in mammalian adipose tissue.

desaturase

the name given to enzymes needed to introduce unsaturation (usually *cis*-olefinic) into acyl chains.

J.L. Harwood Lipid Synthesis and Manufacture (ed. F.D. Gunstone) Sheffield Academic Press, Sheffield (1999) pp.422-466.

desmosterol

the sterol 5,24-cholestadien-3-ol (see structure on next page).

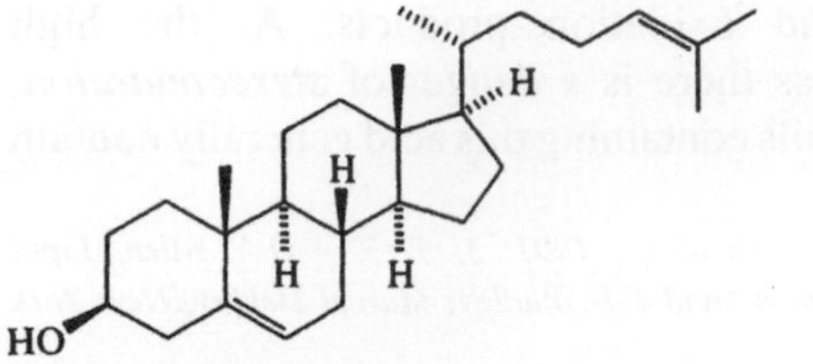

Desmosterol
3β-Cholesta-5,24-dien-3-ol
$C_{27}H_{44}O$
Mol. Wt.: 384.6

detergents

surface-active compounds used for cleaning purposes. These may be *soaps* based on carboxylic acids or they may have other head groups. See also *emulsifiers* and *surfactants*.

Deutsche Gesellschaft fur Fettwissenschaft

Deutsche Gesellschaft fur Fettwissenschaft (DGF). The German society for fat/lipid research. It is responsible for the journal which has had several names including *Fat Science Technology* and, from 1996, was called *Fett/Lipid*. It was changed again in 2000 to *European Journal of Lipid Science and Technology*.

DG

see *diacylglycerols*.

DGDG

see *digalactosyldiacylglycerol*.

DGF

see *Deutsche Gesellschaft fur Fettwissenschaft*.

DHA

see *docosahexaenoic acid*

Dhasco ™

a triacylglycerol oil produced by the alga *Crypthecodinium cohnii* and containing ca 40% of *docosahexaenoic acid*. It is marketed as a source of this acid for nutritional purposes.

D.J. Kyle, *Production and use of a single cell oil which is highly enriched in docosahexaenoic acid*, Lipid Technology, 1996, 8, 107-110.

diacylglycerols

also called diglycerides or DAG. The *sn*–1,2 and *sn*–2,3 compounds are enantiomers (αβ-diacylglycerols) and are isomeric with the *sn*–1,3 compounds (αα-diacylglycerols).The 1,2 (2,3) isomers readily interconvert with the 1,3 compound which is the more stable isomer. 1,2-diacyl-*sn*-glycerols are important intermediates in the biosynthesis and metabolism of triacylglycerols and phospholipids and are vital cellular messengers. They can be synthesised from glycerol or from *monoacylglycerols*, usually with the help of appropriate blocking groups, or from *triacylglycerols* by controlled hydrolysis (*lipolysis*). The melting points of some diacylglycerols are given:

Fatty acid	1,2-isomer	1,3-isomer
Capric	–	44°C
Lauric	20, 39°C	58°C
Myristic	54, 59°C	67°C
Palmitic	50, 66°C	73°C
Stearic	59, 71°C	81°C
Oleic	oil	27°C
Elaidic	ca 25°C	50°C

Many diacylglycerols with two different fatty acids have also been synthesised.

P.E. Sonnet, *Lipid Synthesis and Manufacture* (ed. F.D. Gunstone) Sheffield Academic Press, Sheffield (1999) pp.167-184. R. Bittman, *Lipid Synthesis and Manufacture* (ed. F.D. Gunstone) Sheffield Academic Press, Sheffield (1999) pp.185-207.

| 1,2 isomer | 1,3 isomer | 2,3 isomer |

Diacylglycerols

dialkyl dihexadecylmalonate
malonate esters of the type $R_2C(COOR')_2$ where R is a C_1 to C_{20} alkyl chain and R'OH is a saturated or unsaturated alcohol. These compounds are almost indigestible and can be blended with normal vegetable oils to produce frying oils of reduced energy value.

dibasic acids
generally refers to acids with two carboxyl groups of which the most common have the general structure $HOOC(CH_2)_nCOOH$. They include (value of n in parenthesis): oxalic (0), *malonic* (1), *succinic* (2), *glutaric* (3), *adipic* (4), *suberic* (6), *azelaic* (7), *sebacic* (8), *dodecanedioic* (10), and *brassylic* (11). See also *dimer acids*. Systematic names such as nonanedioic acid indicate the presence of two carboxyl groups and the total number of carbon atoms in each molecule (in this case nine). Some of these are used in the production of polyesters and polyamides.

diesel oil
oils and fats — usually as their methyl or ethyl esters — can be used satisfactorily in diesel engines in complete or partial replacement of the normal diesel fuel. Soybean, rape, palm, animal fats, waste frying oils and others have been used in this way. See *biodiesel*.

diet margarine
see *margarine* and *low-fat spreads*.

differential scanning calorimetry
DSC, technique by which phase transitions on melting or crystallization are recorded as the differential of the heat curve of the sample. This gives a peak, equivalent to a melting point, for the midpoint of a transition.

digalactosyldiacylglycerols
lipid class common in plant membranes. See also *galactosylglycerides*.

Digalactosyldiacylglycerols

digalactosyldiglycerides
see *digalactosyldiacylglycerols.*

digestion
see lipid digestion

diglycerides
see *diacylglycerols.*

diglycerol
see *polyglycerol esters.*

diHETE
dihydroxyeicosatrienoic acids. See *epoxidised arachidonic acids.*

dihomo-γ-linolenic acid
the bis-homologue of γ-linolenic acid (8,11,14-*eicosatrienoic acid*) and an intermediate in the bioconversion of *linoleic acid* to arachidonic acid (*eicosatetraenoic acid*). It can be made chemically from γ-linolenic acid.

dihydroambrettolide
the lactone of 16-hydroxyhexadecanoic acid.

dilatation
the change in volume occurring in a fat on changing from the solid to the liquid state. See *solid fat index.*

dilatometry
the study of dilatation in a dilatometer, a piece of equipment designed to measure volume changes during melting. See *solid fat index.*

dimer acid
dimers of unsaturated acids, mainly C_{36} dibasic acids. They are manufactured under a variety of conditions in the presence of clays or of peroxides. Produced on a commercial scale for use mainly in polyamide resins. They can be reduced to dimer alcohols which are used to prepare urethanes. See also *Guerbet alcohols.*

dimethyloxazolines
4,4-dimethyloxazolines, made from acids or their derivatives and

2-amino-2-methylpropanol, are used for the study of fatty acids by *mass spectrometry.*

F.D. Gunstone, Lipid Synthesis and Manufacture (ed. F.D. Gunstone) Sheffield Academic Press, Sheffield (1999) pp.321-346. W.W. Christie, Beginner's guide to mass spectrometry of fatty acids. General purpose derivatives, Lipid Technology, 1996, 8, 64-66.

dimethyloxazoline derivatives

4,4-dimethyloxazoline derivatives of fatty acids made by heating the acid with excess of 2-amino-2-methylpropanol at 180°C. These derivatives are used in GC–MS studies to determine double bond position in unsaturated acids.

F.D. Gunstone, Lipid Synthesis and Manufacture (ed. F.D. Gunstone) Sheffield Academic Press, Sheffield (1999) pp.321-346.

Dimethyloxazoline derivates
2-alkyl-4,4-dimethyloxazole

dimorphecolic acid

the (*S*)10*t*12*c* form of 9-hydroxy-10,12-octadecadienoic acid is present in *Xeranthemum annum* seed oil. The (*S*)10*t*,12*t* isomer is a component of the seed fat of *Dimorphotheca aurantica*. The acid is isomeric with *coriolic acid* and both are readily dehydrated, especially under acidic conditions, to a mixture of conjugated 8,10,12- and 9,11,13-octadecatrienoic acids.

Dimorphotheca pluvialis

its seed oil contains *dimorphecolic acid* (ca 60%) which is easily dehydrated to a mixture of conjugated triene acids. Attempts are being made to develop this as a commercial crop.

diol lipids

diacyl derivatives of diols such as 1,3-propanediol. Small quantities of such compounds are present in liver, adipose tissue and egg yolk. They are difficult to detect in the presence of the large amounts of triol lipids (glycerol esters).

A. McMordie and M.S. Manku, Combinatorial lipids: new lipid chemical entities, Lipid Technology, 2000, 8-11.

dipalmitoylphosphatidylcholine

DPPC. This simple *phosphatidylcholine* with two identical acyl groups is an important constituent of mammal lung surfactant. It is produced commercially, usually semi-synthetically from lecithin, and utilized in pharmaceutical *liposome* formulations. See *pulmonary surfactant.*

diphosphatidylglycerols
a complex phospholipid class, composed of glycerol (3 mols), phosphoric acid (2 mols), and fatty acids (4 mols). Also called cardiolipin; a major constituent of bacterial membranes but occurs only in mitochondrial membranes in animals.

Diphosphatidylglycerols

directed interesterification
chemically catalysed *interesterification* carried out at a lower temperature (typically 40–60°C) than normal. Reaction is slower at the lower temperature and the equilibrium (randomisation) is disturbed by the crystallisation of the higher melting (more saturated) triacylglycerols.

A. Rozendaal and A.R. Macrae, Lipid Technologies and Applications (ed. F.D. Gunstone and F.B. Padley) Marcel Dekker, New York (1997), pp.223-263.

disialogangliosides
see *gangliosides.*

distillation
a procedure for separating and purifying compounds by converting them to the vapour phase, passing them through a separation column and then condensing them. It can be applied to fatty acids, simple alkyl esters and in some cases to low molecular weight glycerol esters. It is an important industrial process for the isolation of individual saturated acids.

DLMG
dilinoleo mono-γ-linolenoyl glycerol also called Oenotherol ™.

DMAP
see *acylating agent.*

DMOX
see *dimethyloxazolines.*

DMOX derivatives
see *dimethyloxazoline derivatives.*

docosahexaenoic acid
DHA, 22:6 (*n*–3), cervonic acid. This acid is abundant in fish oils (e.g. *tuna oil*) and is a significant component of membrane lipids of most animal tissues especially in lipids of brain, sperm and the retina of the eye. Its presence in human milk fat is considered to be important for the healthy development of infant brain. It attains high levels in some microorganisms (*Crypthecodinium cohnii*, see *Dhasco*™). The level of docosahexaenoic acid in some *fish oils* can be raised by enzymic methods since many lipases discriminate against this acid.

J.-M. Vatele, Lipid Synthesis and Manufacture (ed. F.D. Gunstone) Sheffield Academic Press, Sheffield (1999) pp.1-45. F.D. Gunstone, Enzymes as biocatalysts in the modification of natural lipids, J. Sci. Fd. Agric., 1999, 79, 1535-1549.

docosanoic acid
the saturated C_{22} acid (behenic, mp 80°C) can be obtained by hydrogenation of *erucic acid* and occurs, along with its C_{20} and C_{24} homologues, as a component of *Lophira alata* (15–30%) and *L. procera* (20%) seed fats.

Behenic acid
docosanoic acid
(22:0)
$C_{22}H_{44}O_2$
Mol. Wt.: 340.6

docosapentaenoic acid
DPA, 22:5 (*n*–3), clupanodonic. This *n*–3 isomer (Δ7,10,13,16,19) accompanies docosahexaenoic acid at lower levels in most fish oils. The *n*–6 isomer (Δ4,7,10,13,16) is present in animal tissues.

docosenoic acid

the best known docosenoic acid is the Δ13 isomer (the *cis* isomer, *erucic acid*, melts at 33°C and the *trans* isomer — brassidic acid — melts at 61°C). Erucic acid is present in high-erucic *rapeseed oil* (up to 50%) and in *Crambe oil* (55–60%). Attempts are being made through genetic engineering to develop a rapeseed with >80% of erucic acid. The acid (~20,000 tonnes per year from 57,000 tonnes of erucic oils) is used particularly in the form of its *amide*. Ozonolysis gives the C_{13} dibasic acid, *brassylic acid*. The *cis*-11 isomer (*cetoleic*) is the main 20:1 acid in *fish oils*. The *cis*-5 isomer is present in *meadowfoam oil*. It readily forms a lactone and undergoes other reactions based on the proximity of the double bond to the carboxyl group.

T.A. Isbell, Development of meadowfoam as an industrial crop through novel fatty acid derivatives, Lipid Technology, 1997, 9, 140-144. T.A. Isbell, Lipid Synthesis and Manufacture (ed. F.D. Gunstone) Sheffield Academic Press, Sheffield (1999) pp.401-421. E.C. Leonard, Sources and commercial applications of high-erucic vegetable oils, Lipid Technology, 1994, 6, 79-83.

dodecanedioc acid

the C_{12} dibasic acid, mp 129°C.

Dodecanedioic acid
$C_{12}H_{22}O_4$
Mol. Wt.: 230.3

dodecanoic acid

the C_{12} acid (lauric, mp 44.8°C, bp 130°C/1mm) which is a major acid in *lauric oils* and oils of the *Cuphea* family. *Hydrogenolysis* of this oil gives *dodecanol*.

Lauric acid
dodecanoic acid
(12:0)
$C_{12}H_{24}O_2$
Mol. Wt.: 200.3

dodecanol

the C_{12} alcohol (lauryl alcohol) produced by oligomerisation of ethene (ethylene) or by *hydrogenolysis* of *dodecanoic acid* or ester by the oleochemical industry. It is a valuable *surface-active* compound used mainly as the sulphate ($ROSO_3H$), ethylene oxide adduct ($R(OCH_2CH_2)_nOH$), or ethylene oxide sulphate ($R(OCH_2CH_2)_nOSO_3H$).

9-dodecenoic acid

lauroleic acid, a C_{12} monounsaturated acid.

9Z-Dodecenoic acid
$C_{12}H_{22}O_2$
Mol. Wt.: 198.3

double bond

carbon atoms are usually held together by a two-electron bond (a single bond). When bound by four electrons the linkage is described as a double bond. Compounds containing double bonds are said to be unsaturated and can generally exist in stereoisomeric forms (see *cis* and *trans*). They undergo addition reactions such as *hydrogenation* and are prone to deterioration through *autoxidation*.

double zero

a term to describe rapeseed varieties which produce oil low in *erucic acid* (max 2%) and low in *glucosinolates* (max 30 μmol/g). See also *canola oil* and *rapeseed oil*.

DPA

see *docosapentaenoic acid*.

DPPC

see *dipalmitoylphosphatidylcholine*.

dry fractionation

a procedure in which a completely liquified oil is slowly cooled with gentle stirring. This ensures formation of large crystals in the β or β' form which are more easily and efficiently separated by filtration under reduced pressure in a Florentine filter or under pressure using a membrane filter. See also *fractionation*, *Lanza fractionation*, and *winterisation*.

R.E. Timms, *Lipid Technologies and Applications (ed. F.D. Gunstone and F.B. Padley) Marcel Dekker, New York (1997), pp.199-222. D.A. Allen, Fat modification as a tool for product development. I. Hydrogenation and fractionation, Lipid Technology, 1998, 10, 29-33. E. Deffense, Dry multiple fractionation: trends in products and applications Lipid Technology, 1995, 7, 34-38. T. Willner and K. Weber, High-pressure dry fractionation for confectionery fat production Lipid Technology, 1994, 6, 57-60.*

drying oils

unsaturated oils that dry (harden) on exposure to oxygen through oxidative polymerisation (e.g. *linseed oil, tung oil, dehydrated castor oil*). They are used to produce surface coatings and printing inks.

DSC

see *differential scanning calorimetry.*

E

ECL
see *equivalent chain length.*

ECN
see *equivalent carbon number.*

EET
Epoxyeicosatrienoic acids. See *epoxidised arachidonic acids.*

EFA
see *essential fatty acids.*

egg lecithin
generally refers to *phosphatidylcholines* from egg yolk, sometimes to a lipid extract from the same source which includes other lipid classes (mainly *phosphatidylethanolamines, sphingomyelins, cholesterol* and *triacylglycerols*). It is used as a source of arachidonic acid (*eicosatetraenoic acid*).

egg yolk lipids
about one-third of the weight of the yolk from hen's eggs is contributed by lipids. *Triacylglycerols* and *phosphatidylcholine* are the main components (typically 60% and 25% respectively). *Phosphatidylethanolamine* and *cholesterol* are also present in small amounts (typically 5% each). Minor components are *sphingomyelin* and *plasmalogens.*

eicosanoic acid
the saturated C_{20} acid (arachidic, mp 76.5°C). This acid, along with its C_{22} and C_{24} homologues, is a minor component of *groundnut oil* (total 5-8%). It attains higher levels in some less common seed fats such as rambutan tallow (~35%) and kusum (20–30%).

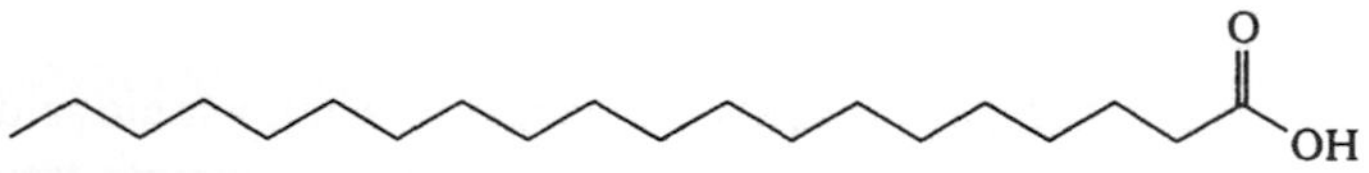

Arachidic acid
eicosanoic acid
(20:0)
$C_{20}H_{40}O_2$
Mol. Wt.: 312.5

eicosanoid cascade
see *prostanoids* and *prostaglandins*.

eicosapentaenoic acid
EPA, 20:5 (*n*–3), timnodonic acid. This acid is present in most fish oils having been obtained from marine algae which are part of the marine food chain, in small amounts in animal phospholipids, and in oils produced by some microorganisms. It is a precursor of some *prostanoids*. See also *docosahexaenoic acid*.

J.-M. Vatèle, Lipid Synthesis and Manufacture (ed. F.D. Gunstone) Sheffield Academic Press, Sheffield (1999) pp.1-45.

eicosatetraenoic acid
(i) the all-*cis n*–6 isomer (5*c*8*c*11*c*14*c*) is better known as arachidonic acid. It is an *essential fatty acid* and a precursor of several important *eicosanoids* such as the *prostaglandins, thromboxanes, leukotrienes* and *lipoxins*. Eicosatetraenoic acid is a minor component of fish oils, but attains a higher level in *animal phospholipids*, and is generally isolated commercially from liver or egg lipids. It is also present in some ferns and can be produced in higher concentrations by fermentation by appropriate microorganisms (see *Arasco*).

J.-M. Vatele, Lipid Synthesis and Manufacture (ed. F.D. Gunstone) Sheffield Academic Press, Sheffield (1999) p.1-45.

Arachidonic acid
5Z,8Z,11Z,14Z-Eicosatetraenoic acid
$C_{20}H_{32}O_2$
Mol. Wt.: 304.5

(ii) The *n*–3 isomer (8*c*11*c*14*c*17*c*) is less well known. It is a member of the *n*–3 family of polyene acids and is present at low levels in fish oils and animal phospholipids.

(iii) The 5*c*11*c*14*c*17*c*-eicosatetraenoic acid is also known as juniperonic acid and is present in many conifer seed oils. The seed oil of the juniper tree (*Juniperus communis*) contains about 18% of this acid which is present almost entirely in the *sn*–3 position of the triacylglycerols. It differs from *eicosapentaenoic acid* in the absence of unsaturation at the 8 position. See also *taxoleic, pinolenic* and *sciadonic acids*.

eicosatrienoic acid
20:3 acids include the 5,8,11 (*Mead's acid*), 7,10,13-, 8,11,14- (*dihomo-γ-linolenic*), 5,11,14-(*podocarpic, pinolenic, sciadonic*) and 11,14,17- isomers. Mead's acid is an *n*–9 acid produced by metabolism of oleic acid, particularly under conditions of EFA deficiency.

The 8,11,14-isomer is the source of some eicosanoids and an intermediate in the biosynthesis of arachidonic acid (*eicosatetraenoic acid*).

The 5,11,14-isomer is present in many conifer seed oils. The seed oils of *Sciadopytis verticillata* contain about 15% of this acid which occurs almost entirely in the *sn*–3 position in the triacylglycerols. It differs from arachidonic acid in that the latter also has an additional double bond in the 8 position. See also *taxoleic* and *juniperonic acids*.

eicosenoic acid
the *cis*-9 (gadoleic), *cis*-11 (gondoic, mp 24–24°C), and *cis*-13 isomers of eicosenoic acid are present in *fish oils*. The *cis*-11 isomer is present in the seed oil of *Camelina sativa* (gold of pleasure) and is a minor component in rapeseed oil.

elaidic acid
trivial name for *trans-9-octadecenoic acid*.

elaido acids
the *trans (E)* isomer of oleic acid is known as elaidic acid, 9*t-octadecenoic acid*. The term elaido is often used to distinguish *trans* isomers from the more common *cis* forms, e.g. linelaidic (9*t*12*t*-octadecadienoic acid) and linolenelaidic (9*t*12*t*15*t*- octadecatrienoic acid).

***Elaies guineensis*, Palmae**
the oil palm, which is the source of *palm oil* (from the mesocarp) and *palm kernel oil* (from the kernel). Mainly grown in South East Asia and Africa. See *Elaeis olifera*.

***Elaies olifera*, Palmae**
the South American oil palm. See *Elaeis guineensis*.

eleostearic acid
see *conjugated unsaturation*.

elongase
the name given to enzymes required for chain-elongation, usually by a two-carbon unit based on acetate or malonate. Similar propionate or methylmalonate derivatives lead to C_3 extensions and usually give methyl-branched acids. See *fatty acids synthetase*.

J.L. Harwood, Lipid Synthesis and Manufacture (ed. F.D. Gunstone) Sheffield Academic Press, Sheffield (1999) pp.422-466.

ELSD
see *evaporative light-scattering detector.*

emulsifier
a surface-active compound which promotes the formation of emulsions between water and fatty or oily compounds. Important in cleaning processes and also in many foods. See also food emulsions. Food emulsifiers are important in reducing dietary energy values in that with their assistance it is possible to make spreads with less fat per unit weight (and more water). Such spreads have lower energy values than full fat products on a weight for weight basis.

N. Krog, Lipid Technologies and Applications (ed. F.D. Gunstone and F.B. Padley) Marcel Dekker, New York (1997), pp.521-534. C.E. Stauffer, Bailey's Industrial Oil and Fat Products, (ed. Y.H. Hui) John Wiley & Sons, New York (1996), Volume 3, pp.483-522. D.J. McClements, Food Lipids: Chemistry, Nutrition, and Biotechnology, (ed. C.C. Akoh and D.B. Min) Marcel Dekker, New York (1998) pp.55-88. G.L. Hasenhuettl, Lipid Synthesis and Manufacture (ed. F.D. Gunstone) Sheffield Academic Press, Sheffield (1999) pp.371-400.

enanthic acid
the trivial name for heptanoic acid (7:0).

enzymic enhancement
a procedure by which mixtures of free acids or glycerol esters are enhanced in one or more of the fatty acid components by reaction with an enzyme which discriminates against the acid(s) in question. The procedure is applied particularly to acids which have a double bond close to the carboxyl group such as *docosahexaenoic, eicosapentaenoic, arachidonic* and *γ-linolenic acid.* See *borage seed oil* and *evening primrose oil,*

F.D. Gunstone, Progress in Lipid Research, 1998, 37, 277-305. G.P. McNeill, Lipid Synthesis and Manufacture (ed. F.D. Gunstone) Sheffield Academic Press, Sheffield (1999) pp.288-320. F.D. Gunstone, Enzymes as biocatalysts in the modification of natural lipids, J. Sci. Fd. Agric., 1999, 79, 1535-1549.

EPA
see *eicosapentaenoic acid*

EpET

epoxyeicosatrienoic acid. See *epoxidised arachidonic acid*.

epichlorohydrin

(2,3-epoxypropyl chloride) a synthetic compound much used to prepare glycerol esters and ethers.

P.E. Sonnet, Lipid Synthesis and Manufacture (ed. F.D. Gunstone) Sheffield Academic Press, Sheffield (1999) pp.162-184. R. Bittman, Lipid Synthesis and Manufacture (ed. F.D. Gunstone) Sheffield Academic Press, Sheffield (1999) pp.185-207.

Epichlorohydrin
2,3-epoxypropylchloride
C_3H_5ClO
Mol. Wt.: 92.5

epoxidation

the reaction by which olefinic acids are converted to *epoxy acids*. This is a *cis* addition of oxygen to the double bond, usually effected by a *peroxy acid* such as peroxyformic or peroxyacetic. The reaction is carried out on an industrial scale to produce *epoxidised soybean oil, epoxidised linseed oil* etc. These are used as plasticizers–stabilizers for polyvinylchloride (PVC).

F.D. Gunstone, Lipid Technologies and Applications (ed. F.D. Gunstone and F.B. Padley) Marcel Dekker, New York (1997), pp.759-769. M. Rusch and S. Warwel, Peroxy fatty acids: lipase-catalysed preparation and epoxidation, Lipid Technology, 1996, 8, 77-80,

epoxidised arachidonic acids

enzymic *epoxidation* of arachidonic acid (*eicosatetraenoic acid*) gives four monoepoxyeicosatrienoic acids (EET or EpET) in each of which one double bond has been epoxidized. Further epoxidation gives a range of diepoxyeicosadienoic acids whilst hydrolysis of the monoepoxy acids give dihydroxyeicosatrienoic acids (DiHETE).

epoxy acids

epoxy acids are made by epoxidation of olefinic acids. They also occur naturally as glycerol esters (e.g. *vernolic, coronaric, alchornoic*). See also *cutins* and *furanoid* acids. Attempts are being made to develop *Vernonia galamensis* and *Euphorbia lagascae* as commercial crops. Both are rich in *vernolic acid.*

F.D. Gunstone, Lipid Technologies and Applications (ed. F.D. Gunstone and F.B. Padley) Marcel Dekker, New York (1997), pp.759-769.

epoxy oils
epoxidised soybean oil and epoxidised linseed oil, used as plasticizers and stabilizers in polymers especially PVC, are made by reaction of the oil with peroxyformic or peroxyacetic acid. In this reaction all the double bonds are epoxidised.

F.D. Gunstone, Lipid Technologies and Applications (ed. F.D. Gunstone and F.B. Padley) Marcel Dekker, New York (1997), pp.759-769.

epoxystearic acids
cis and *trans* 9,10-epoxystearic acids (mp 59.5–59.8°C and 56–57°C respectively) are formed by *epoxidation* of *oleic* and *elaidic acids* respectively.

equivalent carbon number
ECN; chromatographic retention behaviour of triacylglycerols relative to the trisaturated homologues, which are defined to have ECN equal to their carbon numbers, i.e. the total number of acyl carbon atoms excluding the glycerol carbon atoms. It is also applied to *molecular species* of other *lipid classes* and is used in connection with *high-performance liquid chromatography* and high-temperature *gas chromatography*.

equivalent chain length
ECL; chromatographic retention behaviour of fatty acid methyl esters relative to the saturated homologues, which exhibit a straight line when the log of the retention times are plotted against the number of acyl carbons atoms. Used in the tentative identification of *fatty acids* by *gas chromatography*.

ergosterol
a 24β-methyl sterol.

| **Ergosterol** | (3β,22E)-Ergosta-5,7,22-trien-3-ol |
| $C_{28}H_{44}O$ | Mol. Wt.: 396.6 |

erucic acid

trivial name for docosenoic acid (13c-22:1, mp 33.5°C). It is an important acid which occurs in seed oils of the Cruciferae, e.g. *rape, mustard and crambe*. Modern varieties of rapeseed oils have been bred to contain less than 2% erucic acid compared to the original 30–50% since this acid may have harmful nutritional properties. Oils with high levels of erucic acid are also being developed since erucic acid is used for oleochemical purposes, e.g. as its amide.

C. Leonard, Sources and commercial applications of high-erucic vegetable oils, Lipid Technology, 1994, 6, 79-83.

essential fatty acids

polyunsaturated acids of the *(n–6)* and *(n–3) families* which are essential for life and good health. They cannot be biosynthesised by animals and they (or some suitable precursor) must be obtained from plant sources as part of the diet.

esterification

the reaction by which esters are formed from alcohols and acids, usually in the presence of an acidic catalyst, or with the more reactive acid anhydrides or chlorides for which no catalyst is required. Esters can also be changed to other esters by *alcoholysis, acidolysis* and *interesterification.*

W.W. Christie, Advances in Lipid Methodology — Two (ed. W. W. Christie) The Oily Press, Dundee (1993) pp.69-111.

esterified propoxylated glycerols

esters of propoxylated glycerol and long-chain acids. Glycerol is reacted with propylene oxide to give a trihydric alcohol which can be acylated with long-chain fatty acids. The esters resist lipolysis and have zero energy value. The products are thermally stable and suitable for use in baking and frying applications.

estolides

these are dimers (or oligomers) consisting of an ester between an acid and a hydroxy acid. They may occur naturally (as in *stillingia oil*) or be formed along with *dimer acids.* A typical monoestolide from oleic acid would have a structure such as that shown on the following page.

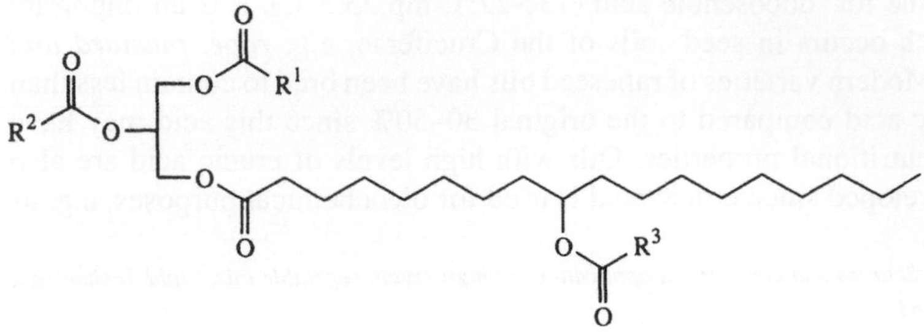

Typical estolide

ethanoic acid
see *acetic acid.*

ether lipids
usually derivatives of the type $ROCH_2CHOHCH_2OH$ where R is a straight or branched, alkyl or alkenyl chain. Such compounds may be natural as in shark liver oil or other marine oils where they generally occur as alkyl acyl, alkenyl acyl or diacyl derivatives. Synthetic compounds are generally made from *epichlorohydrin.* See also *batyl alcohol, chimyl alcohol, plasmalogens* and *platelet activating factor.*
K. Urata et al., *Ether lipids based on the glyceryl ether skeleton: present state, future potential,,* J. Am. Oil Chem. Soc., 1996, 73, 818-830.

ethoxylation
a reaction of alcohols (or carboxylic acids or amines) with ethylene oxide.

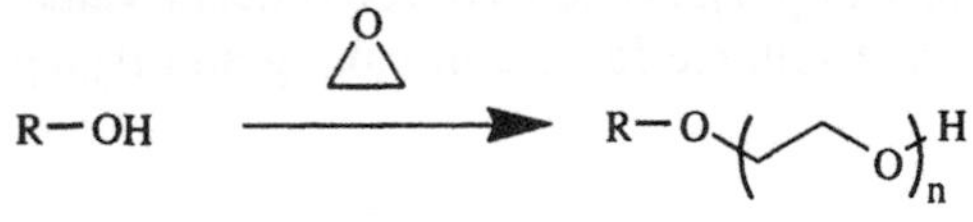

Ethoxylation

ethoxyquin
a dihydroquinoline derivative used as an antioxidant. It is frequently added to fish meal.

Ethoxyquin
6-Ethoxy-1,2-dihydro-2,2,4-trimethylquinoline
$C_{14}H_{19}NO$
Mol. Wt.: 217.3

Eurolipid
an association of European national organisations devoted to the study of lipids.

European Journal of Lipid Science and Technology
A new name for the journal *Fett/Lipid* starting in 2000. It is the official journal of the *Deutsche Gesellschaft fur Fettwissenschaft*.

evaporative light scattering detector
sometimes termed mass detector; a type of detector used for liquid chromatography, e.g. high-performance liquid chromatography and gel-permeation chromatography. The detector is based on the principle of light-scattering of the nebulized effluent from the columns with evaporation of the solvents. These instruments have found special applications in lipid analysis as universal detectors, since the response is independent of the nature of the mobile phase.
W.W. Christie, Advances in Lipid Methodology — One (ed. W. W. Christie) The Oily Press, Ayr (1992) p.239.

evening primrose oil
oil obtained from *Oenothera biennis* or *O. lamarkiana*. This oil is used as a dietary supplement or as a pharmaceutical because it contains *γ-linolenic acid* (8–10%) as well as a high level of linoleic acid (~70%). Other conventional sources of GLA are *borage oil* and *blackcurrant seed oil*. The level of γ-linolenic acid in evening primrose oil can be raised to 20–25% by *enzymic enhancement*.
F.D. Gunstone, Progress in Lipid Research, 1992, 31, 145. D.F. Horrobin, Progress in Lipid Research, 1992, 31, 163.

exocarpic acid

an acetylenic acid with conjugated unsaturation (9a11a13t-18:3) present in isano (boleko) oil.

extraction

(i) industrial procedures by which oils and fats are recovered from vegetable or animal sources by *pressing, solvent extraction* or *rendering*.

M.A. Williams, Lipid Technologies and Applications (ed. F.D. Gunstone and F.B. Padley) Marcel Dekker, New York (1997), p.113.

(ii) laboratory procedures by which lipids are obtained from natural samples and from foods. Dry seeds can be extracted with hexane in a Soxhlet apparatus after crushing. Wet tissue can be homogenised in a chloroform/methanol/water mixture according to the method of *Folch et al.* or of *Bligh and Dyer*. The extraction process of Rose and Gottlieb is useful for foods which also contain proteins and carbohydrates.

W.W. Christie, Advances in Lipid Methodology — One (ed. W. W. Christie) The Oily Press, Ayr (1992) p.1-17. W.W. Christie, Advances in Lipid Methodology — Two (ed. W. W. Christie) The Oily Press, Dundee (1993) p.195-213. F. Shahidi and J.P.D. Wanasundare, Food Lipids: Chemistry, Nutrition, and Biotechnology, (ed. C.C. Akoh and D.B. Min) Marcel Dekker, New York (1998) p.115-136.

F

Fabry's disease
malfunctioning of the enzymes responsible for the stepwise breakdown of ceramide trihexosides to ceramides is the cause of Fabry's disease (β-galactosidase) and of Gaucher's disease (β-glucosidase).

false flax
see *Camelina sativa.*

FAME
see fatty acid methyl esters.

Fantesk™
a stable emulsion of oil, water, and carbohydrate in the form of a soft gel or a free-flowing powder which can be used in low-fat foods or other applications.
K. Eskins and G. Fanta, Fantesk: carbohydrate-oil composites useful in low-fat foods, cosmetics, drugs, and industrial applications Lipid Technology, 1996, 8, 53-57.

Farber's disease
a rare lipidosis resulting from malfunction of the enzyme (ceramidase) responsible for breakdown of ceramides to sphingosines and fatty acids. It results in the accumulation of ceramides.

fat mimetics
see *structured fats.*

fats
the bulk storage material produced by plants, animals and microorganisms that contains aliphatic moieties, such as fatty acid derivatives. These are mainly, but not entirely, mixtures of *triacylglycerols* (triglycerides) and are known as oils or fats depending on whether they are liquid or solid at room temperature. See also *butter.*

Fat Science Technology
see *European Journal of Lipid Science and Technology.*

fat splitting
fats can be hydrolysed to free acids by water itself in what is probably a homogeneous reaction between fat and water dissolved in the oil phase. This

is usually effected in a continuous high-pressure, uncatalysed, countercurrent process at 20–60 bar and 250°C. At these high temperatures the products may be discoloured and both the fatty acid and the glycerol may be subsequently distilled.

fat-soluble vitamins
see *vitamins*.

fat substitutes
see *structured fats*.

fatty acid methyl esters
the derivatives most used for gas chromatography. They are readily made from lipids by *methanolysis*.
W.W. Christie, Advances in Lipid Methodology — Two (ed. W.W. Christie) The Oily Press, Dundee (1993) pp.69-111. F.D. Gunstone, Lipid Synthesis and Manufacture (ed. F.D. Gunstone) Sheffield Academic Press, Sheffield (1999) pp.321-346.

fatty acids
alkanoic and alkenoic acids; these are *saturated* or *unsaturated* organic acids generally having an unbranched chain with an even number of carbon atoms. They are major components of most lipids. See also individual acids (e.g. *stearic, oleic*) and classes of acids (e.g. *hydroxy acids*).

$CH_3(CH_2)_n(CH=CHCH_2)_m(CH_2)_pCOOH$

General formula for typical fatty acids
(saturated: m = 0, monounsaturated: m = 1, polyunsaturated: m > 1)

fatty acids synthetase
a group of enzymes which are involved in the *de novo* formation of fatty acids in animals and plants, but also chain lengthening of, for example, fatty acids from the diet (*elongases*).
C.F. Semenkovich, Regulation of fatty acid synthetase (FAS), Progress in Lipid Research, 1999, 36, 1, 45-35

fatty alcohols
medium-chain and long-chain alcohols related to the *fatty acids*. They occur naturally in *wax esters* and are produced industrially by reduction (*hydrogenolysis*) of acids or methyl esters. They are used as *surfactants*, usually as sulphate or polyoxyethylene derivatives.

$CH_3(CH_2)_n(CH=CHCH_2)_m(CH_2)_pOH$

General formula for typical fatty alcohols
(saturated: m = 0, monounsaturated: m = 1, polyunsaturated: m > 1)

fatty aldehydes
these are medium and long-chain aldehydes related to the fatty acids. They are sometimes present in waxes and are produced on chemical hydrolysis of *plasmalogens* where they are bound in the form of their enol ethers.

$CH_3(CH_2)_n(CH=CHCH_2)_m(CH_2)_pCHO$

General formula for typical fatty aldehydes
(saturated: m = 0, monounsaturated: m = 1, polyunsaturated: m > 1)

FCL
see *fractional chain length*.

FEDIOL
a group of European professional associations or syndicates of the (fatty) oil industry based in Brussels.

ferulic acid
4-hydroxy-3-methoxycinnamic acid. This is widely distributed in plants, often as an ester. It shows marked antioxidant activity. See *rice bran oil*.

Fett/Lipid
a journal devoted to oils and fats and published by the *Deutsche Gesellschaft fur Fettwissenschaft*. Previously called *Fett Wissenschaft Technologie* and *Fette Seifen Anstrichmittel*, its name was changed to *Fett/Lipid* in 1996 and changed again in 2000 to *European Journal of Lipid Science and Technology*.

Fette Seifen Anstrichmittel
see *European Journal of Lipid Science and Technology*.

Fett Wissenschaft Technologie
see *European Journal of Lipid Science and Technology*.

FFA
see *free fatty acids*.

FID
see *flame-ionisation detector*.

filled milk
a milk-like product in which butterfat has been replaced by a vegetable oil or fat.

fish oil
lipid extracted from the body, muscle, liver, or other organ of fish. The major producing countries are Japan, Chile, Peru, Denmark and Norway and the main fish sources are *herring, menhaden, capelin, anchovy, sand eel, sardine, tuna* and *cod* (liver). In the five-year period 1996–2000, production of fish oil averaged 1.11 million tonnes a year. Fish oils contain a wide range of fatty acids from C_{14} to C_{26} in chain length with 0–6 double bonds. The major acids include saturated (14:0, 16:0, and 18:0), monounsaturated (16:1, 18:1, 20:1, and 22:1) and *n*–3 polyene members (18:4, 20:5, 22:5, and 22:6). Fish oils are easily oxidized and are commonly used in *fat spreads* only after partial *hydrogenation*. However, they are the most readily available source of *n*–3 polyene acids, especially *eicosapentaenoic acid* and *docosahexaenoic acid*, and with appropriate refining procedures and antioxidant addition these acids can be conserved and made available for use in food. The long-chain polyene acids are valuable dietary materials and there is a growing demand for high quality oil rich in *eicosapentaenoic acid* and *docosahexaenoic acid*. The proportion of these acids can be raised by *enzymic enhancement*. The *n*–3 (omega-3) oils and their concentrates are available in capsule form or can be incorporated directly into baking and spreading fats. Protein-rich meal remaining after oil extraction is valuable feed for land animals and farmed fish.

C.F. Moffat, Fish oil triglycerides: a wealth of variations, Lipid Technology, 1995,7, 125-129.
G.M. Pigott, Bailey's Industrial Oil and Fat Products, (ed. Y.H. Hui) John Wiley & Sons, New York (1996), Volume 3, pp.225-254. S. Madsen, Develeopment of a reduced-fat spread enriched with long-cahin n–3 fatty acids, Lipid Technology, 1998, 10, 129-132.

fixed oils
pharmacopoeial term for oils and fats of defined origins.

flame ionisation detector
FID; the most universal detector for *gas chromatography* commonly used for the analysis of lipids such as fatty acid methyl esters, sterols and triacylglycerols. The detector measures ions that are generated when organic compounds are combusted. It can be used for virtually all organic compounds

and has high sensitivity and stability, a low dead volume, a fast response time and the response is linear over a wide range.

flax seed oil
see *linseed oil.*

Florisil™
trade name for an *adsorbent* used for column chromatography of lipids. It is a mixture of magnesia and silica. See also *adsorption chromatography.*

fluoro acids
16-fluoropalmitic acid and 18-fluoro-oleic acid among others, present in the seed oil of the West African *Dichapetalum toxicarium* They are responsible for the toxic nature of these seeds. Metabolism leads eventually to the very toxic fluoroacetic acid.

Folch extraction
In the Folch extraction procedure ground or homogenized tissue is shaken with a 2:1 mixture of chloroform and methanol and the organic extract is subsequently partitioned with aqueous potassium chloride solution. It is important that the ratio of chloroform, methanol and water in the final mixture be 8:4:3. The extracted lipids are in the (lower) chloroform layer.
W.W. Christie, Advances in Lipid Methodology – Two, The Oily Press, Dundee, 1993, pp.195-213.

food emulsions
a wide variety of emulsions exist in food. Examples include *margarine, butter* and mayonnaise (semi-solid; water-in-oil emulsions), milk and dressings (liquid; oil-in-water emulsions) and ice cream (mixture).

formic acid
the first member of the homologous series of *alkanoic acids.* Its systematic name, indicating that it is the C_1 acid, is methanoic acid.

$$\underset{H}{\overset{O}{\|}}\diagup C \diagdown OH$$

Formic acid
methanoic acid
(1:0)
CH_2O_2
Mol. Wt.: 46.0

FOSFA

Federation of Oil, Seeds and Fats Associations producing rules for the trading of oils.

Fourier-transform infrared spectroscopy

this improved computer-aided procedure for measuring and recording infrared spectra makes it possible to measure several parameters without use of solvents and avoids laborious titration procedures. This includes *iodine value, saponification value, acid value, peroxide value, anisidine value*, and measurement of *trans acids*. Linked to a gas chromatographic system, it is of great value for characterisation of *trans* acid.

F.R. van de Voort et al., Moving FTIR spectroscopy into the quality control laboratory, Lipid Technology, 1996, 8, 89-91 and 117-119. J. Sedman et al., New Technologies and Applications in Lipid Analysis (eds R.E. McDonald and M.M. Mossoba) AOCS Press, Champaign, Illinois, 1997, 283-324.

fractional chain length

FCL. By definition the *equivalent chain length* (ECL) of a saturated acid/ester is equal to the number of carbon chains in the acid. Thus, for stearic esters it is 18.00. The ECL of oleic acid and its esters is a non-integral number where the actual value depends on the chromatographic conditions, especially the nature of the stationary phase. For example it may be 18.53 on one polar stationary phase. The value +0.53 is then the FCL corresponding to a selected stationary phase and a particular structural unit, in this case a *cis* double bond in the $\Delta9/\omega9$ position.

fractionation

a procedure for separating oils and fats on a commercial scale into two or more components. This involves two steps: establishment of the crystallisation equilibrium followed by separation of the solid and liquid fractions by reduced-pressure or high-pressure filtration. In this way a solid fraction of higher melting point and lower solubility (the *stearin*) is separated from the liquid fraction of lower melting point and higher solubility (the *olein*). These products extend the range of use of the original oil or fat. The procedure is applied particularly to *palm oil, palm kernel oil*, and *anhydrous milk fat*. In *dry fractionation* the stearin crystallises from the liquid oil; in *solvent fractionation* an appropriate solvent such as acetone is used.

R.E. Timms, Lipid Technologies and Applications (ed. F.D. Gunstone and F.B. Padley) Marcel Dekker, New York (1997), pp.199-222. D.A. Allen, Fat modification as a tool for product development, Lipid Technology, 1998, 10, 29-33. E. Deffense, Dry multiple fractionation: trends in products and applications, Lipid Technology, 1995, 7, 34-38. T. Willner and K. Weber,

High-pressure dry fractionation for confectionery fat production, Lipid Technology, 1994, 6, 57-60. M. Kellens, Edible Oil Processing (eds W. Hamm and R.J. Hamilton) Sheffield Academic Press, Sheffield (2000), pp.156-168.

free (unesterified) fatty acids

fatty acids in unbound (unesterified) form. *Oils* and *fats* which are mainly *triacylglycerols* contain in their natural (crude) state small amounts of free acids which may be removed by processing. See *neutralization, physical refining*; also *acid value*. It is a fraction with an important physiological function in animal and plant tissues.

frying fats

these are fats used for cooking by the process of frying. The hot fat acts as a heat transfer agent. Some of it remains in the fried food. During frying the heated oil may undergo several undesirable changes including *hydrolysis, autoxidation, stereomutation*, cyclisation, and polymerisation.

T.L. Mounts, Lipid Technologies and Applications (ed. F.D. Gunstone and F.B. Padley) Marcel Dekker, New York (1997), pp.433-451. M. M. Blumenthal, Bailey's Industrial Oil and Fat Products, (ed. Y.H. Hui) John Wiley & Sons, New York (1996), Volume 3, pp.429-481. K. Warner, Food Lipids: Chemistry, Nutrition, and Biotechnology, (ed. C.C. Akoh and D.B. Min) Marcel Dekker, New York (1998) pp.167-180. E.G. Perkins and M.D. Erickson (ed), Deep frying — chemistry, nutrition, and practical applications, AOCS Press, Champaign, Illinois, 1996.

FSA, FST

see *European Journal of Lipid Science and Technology*.

FTIR

see *Fourier-transform infrared spectroscopy*.

fucosterol

a sterol similar to cholesterol but with an ethylidene substituent in the side-chain.

Fucosterol
(3β,24E)-Stigmasta-5,24(28)-dien-3-ol
$C_{29}H_{48}O$
Mol. Wt.: 412.7

Fullers earth

see *bleaching earth*.

furanoid acids

fatty acids containing a furan ring. Furanoid fatty acids are a rare group of fatty acids found in some plants and as minor components of many fish oils (typical structure below). Furanoid acids of lower molecular weight, *urofuranic acids*, have been recognized in urine and blood.

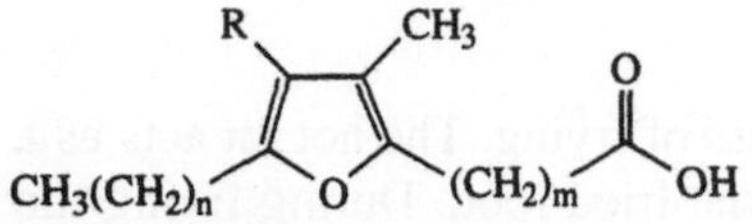

Furanoid acids
R = H or CH$_3$

G

gadelaidic acid
trivial name for *trans*-9-eicosenoic acid (*9t*-20:1, mp 54°C).

gadoleic acid
trivial name for *cis*-9-eicosenoic acid (*9c*-20:1, mp 23–23.5°C) present in many fish oils.

gaidic acid
trivial name for 2-hexadecenoic acid, mp 40.5–41.7°C (*cis* isomer) and 46.5–47.6°C (*trans* isomer).

galactolipids
lipids containing one or more galactose units. See *mono-* and *digalactosyldiacylglycerols, glycosylceramide,* and *galactosylglycerides.*

galactosylceramides
see *glycosylceramides.*

galactosylglycerides
diacylglycerols with one to four (commonly one or two) galactose units attached to the *sn*–3 position by glycosidic bonds. Common membrane lipid components of plants (chloroplasts). See *galactolipids, glycosylglycerides.*

gangliosides
trivial name for a family of glycosphingolipids with a *ceramide* linked to a carbohydrate moiety with *sialic acid* units. Occurs in the gray matter of brain, spleen, erythrocytes and many other animal tissues. Divided into three main classes, monosialogangliosides, G_{M1}, G_{M2} (tay-Sachs ganglioside), G_{M3} (haematoside); disialogangliosides, G_{D1a}, G_{D1b}; and trisialogangliosides, G_{T1}.
K.-J. Jung and R.R. Schmidt, *Lipid Synthesis and Manufacture (ed. F.D. Gunstone) Sheffield Academic Press, Sheffield (1999) pp.208-249.*

Garcinia indica
this seed fat (kokum butter) from India is rich in stearic (~53%) and oleic (~40%) acids and contains ~74% of StOSt in its glycerol esters. It can be used as a *cocoa butter equivalent.*

gas chromatography
GC, generally *chromatography* with a carrier gas (hydrogen, helium or nitrogen) as the mobile phase. Useful for any volatile lipid compound, such as fatty acid *methyl esters, triacylglycerols* and *sterol esters,* or compounds that can be made volatile, such as phospholipids after (enzymic) dephosphorylation.
W.W. Christie, Gas Chromatography and Lipids — A Practical Guide, The Oily Press, 1989. K. Eder, Review: Gas chromatographic analysis of fatty acid methyl esters, Journal of Chromatography B, 671, 1995, 113-131.

Gaucher's disease
see *Fabry's disease.*

GC
see *gas chromatography.*

G$_{D1a}$, G$_{D1b}$
see *gangliosides.*

geddic (gheddic) acid
trivial names for tetratriacontanoic acid (34:0, mp 34°C).

genetic modification
a form of biotechnology by which existing genes are modified or new genes are introduced into living systems. Plants can be modified to show novel traits of agricultural significance. These include herbicide tolerance, male sterility, and resistance to insects, virus, fungi and bacteria. Modification can also lead to the production of oils with changed fatty acid and triacylglycerol composition. See *lauric-canola.*
M. Lassner, Transgenic oilseed crops: a transition from basic research to product development, Lipid Technology, 1997, 9, 5-9. V.C. Knauf and A.J. Del Vecchio, Food Lipids: Chemistry, Nutrition, and Biotechnology, (ed. C.C. Akoh and D.B. Min) Marcel Dekker, New York (1998) pp.779-805.

Geotrichum
see *lipases.*

ghee
a solid fat-based product made in India from cow or buffalo ripened milk. It is less perishable than butter and therefore more suitable for a tropical climate.

K.T. Achaya, Lipid Technologies and Applications (ed. F.D. Gunstone and F.B. Padley) Marcel Dekker, New York (1997), pp.369-390.

GLA
γ-linolenic acid, see *octadecatrienoic acid.*

GLC
see *gas-liquid chromatography.*

globosides
see *glycosylceramides.*

glucosinolates
undesirable sulphur-containing compounds present in rape seed which remain in the seed meal. They have been reduced to an acceptable level in the double-zero varieties.

glutaric acid
the C_5 dibasic acid (mp 97°C).

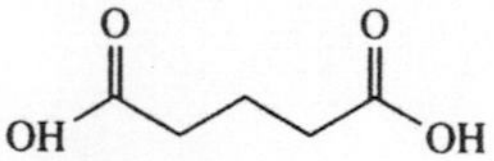

Glutaric acid
Pentanedioic acid
$C_5H_8O_4$
Mol. Wt.: 132,1

glycerol
trivial name for 1,2,3-trihydroxypropane. The most common natural carrier for acyl groups and the basis of many *lipid classes,* e.g. *triacylglycerols* and *glycerophospholipids.* Glycerol is liberated during commercial production of *fatty acids, soaps, esters,* etc. About one million tonnes is produced each year by this natural route and also by production from petrochemicals (propene). It has many uses.

glycerol 1, 3-dioleate 2-palmitate
the glycerol ester (*OPO, Betapol™*) made by reaction of *tripalmitin* with oleic-rich sunflower oil in the presence of a 1,3-regiospecific enzyme. Reaction is confined to the *sn*−1 and 3 positions. This triacylglycerol is rare in nature but is present in human milk fat.

R.G. Jensen, Human milk lipids as a model for infant formulas, Lipid Technology, 1998, 10, 34-38. G.P. McNeill, Lipid Synthesis and Manufacture (ed. F.D. Gunstone) Sheffield Academic Press, Sheffield (1999) pp.288-320.

glycerolysis

the reaction of triacylglycerols with glycerol in the presence of an alkaline or enzymic catalyst to produce a mixture of mono-, di- and tri-acylglycerols. This reaction is used to prepare mono- and diacylglycerols on an industrial scale. These can be separated by distillation.

triacylglycerol + glycerol ⇆ monoacylglycerol + diacylglycerol

N. Krog, Lipid Technologies and Applications (ed. F.D. Gunstone and F.B. Padley) Marcel Dekker, New York (1997), pp.521-534. G.P. McNeill, Lipid Synthesis and Manufacture (ed. F.D. Gunstone) Sheffield Academic Press, Sheffield (1999) pp.288-320 .

glycerophospholipids

lipids based on *glycerol* containing a phosphate ester at the *sn*–3 position and acyl groups in the *sn*–1 and *sn*–2 positions. The main classes are *phosphatidylcholine, phosphatidylethanolamine, phosphatidylglycerol, phosphatidylinositol, phosphatidylserine* and *diphosphatidylglycerol.*

Glycine max (Leguminosae)

see *soybean oil.*

glycolipids

general name for all lipids linked to any type of carbohydrate moiety. The main plant glycolipids are *mono-* and *digalactosyldiacylglycerols* though *sterol glycosides* and *cerebrosides* also exist The animal kingdom contains mainly *glycosphingolipids,* e.g. *gangliosides and cerebrosides.*

glycosphingolipids

see *glycolipids.*

glycosyl ceramides

ceramide hexosides; the simplest types of glycosphingolipids with up to six (or more) monosaccharide units, usually galactose and/or glucose. Trivial names are used for monoglycosylceramides (*cerebrosides*), diglycosylceramides (cytosides) and tetraglycosyl-ceramides (globosides).

glycosylglycerides

compounds in which one to four monosaccharide units are linked glycosidically to 1,2-diacyl-*sn*-glycerol. Present in the photosynthetic membranes of plants. See also *galactosylglycerides*.

E. Heinz, Plant glycolipids: structure, isolation and analysis, Advances in Lipid Methodology — Three (ed. W.W. Christie), The Oily Press, Dundee, (1996), pp.211-332.

glyoxisomes
see *peroxisomes*.

G$_{M1}$, G$_{M2}$, G$_{M3}$
see *gangliosides*.

GMS
glycerol monostearate, see *monostearin*.

gold of pleasure
see *Camelina sativa*.

gondoic acid
see *eicosenoic acids*.

Good-Fry oil
A blend of high-oleic vegetable oil (corn or sunflower) with up to 6% of *sesame* and/or *rice bran oil*, both of which show high oxidative stability by virtue of the *antioxidants* among their minor components.

gorlic acid
13-(2-cyclopentenyl)-6-tridecenoic acid, see *cyclopentenyl acids*.

Gorlic acid
13-(2-cyclopenten-1*R*-yl)-6*Z*-tridecenoic acid
C$_{18}$H$_{30}$O$_2$
Mol. Wt.: 278.4

Gossypium barbadense (Malvacaea)
see *cottonseed oil*.

Gossypium herbaceum (Malvaceae)
see *cottonseed oil*.

Gossypium hirsutum (Malvaceae)
see *cottonseed oil*.

gossypol
complex phenolic substance occurring in crude *cottonseed oil* and removed
by the refining process.

gourmet oils
another name for *speciality oils*. These are generally minor oils available in
small quantities. They are used as food oils but also in cosmetics, toiletries,
and with pharmaceuticals.

grapeseed oil
a linoleic-rich oil (ca 75%) from *Vitis vinifera*. See also *speciality oils*.

GRAS

acronym for "generally recognised as safe". A description given to some
materials by the Food & Drug Administration of the USA.

Grasas y Aceitas
Spanish journal devoted to lipid research and published by the Instituta de la
Grasa, Seville, Spain.

groundnut oil
seed oil of the legume *Arachis hypogea*, also known as peanut oil, monkeynut
oil, or arachis oil. World production of the oil is about 4.6 million tonnes a
year. It is grown widely, but especially in India, China, and the USA. Its major
acids are palmitic (8–14), oleic (36–67), and linoleic (14–44%) with typical
values of 13, 37, and 44 along with 5–8% (total) of C_{20}, C_{22}, and C_{24} saturated
and monoene acids. The major triacylglycerols of one sample of its seed oil
were reported to be LLL (6), LLO (26), LLS (8), LOO (21), LOS (13),
OOO (5), OOS (16), and others (5%). There is also a high-oleic variety.
Groundnut oil shows good oxidative stability and is considered to have a
desirable flavour.
C.T. Young, Bailey's Industrial Oil and Fat Products, (ed. Y. H. Hui) John Wiley & Sons, New York
(1996), Volume 2, pp.377-392.

G_{T1}
see *gangliosides*.

Guerbet acids

see *Guerbet alcohols.*

Guerbet alcohols

monofunctional dimeric alcohols made by heating saturated alcohols at 200–300°C with potassium hydroxide or potassium alkoxide. Reaction involves aldol dimerisation of the corresponding aldehyde. The branched chain alcohol can be oxidised to the corresponding acid (Guerbet acids). These compounds have depressed melting points (by reason of their branching) and good oxidative stability and are used in cosmetics and as plasticisers and lubricants.

Fatty Alcohols, Raw materials, Uses; Henkel KgaA, Düsseldorf, Germany, 1981.

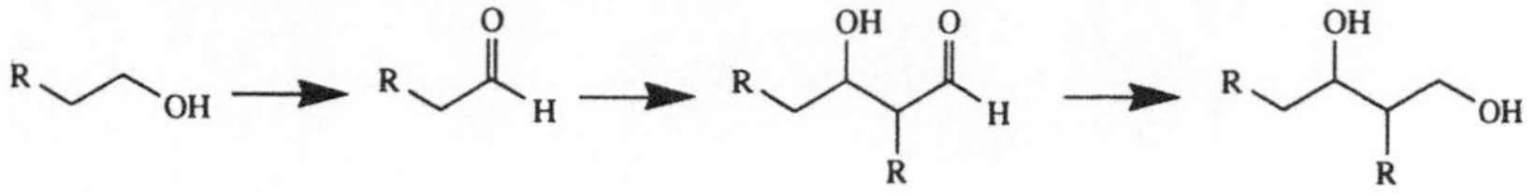

Guerbet alcohols

H

haematoside

GM_3; a monosialoganglioside. Present in white and red blood cells. See *gangliosides*.

Halphen test

a colorimetric test for *cyclopropene acids*. When esters of such acids react with a solution of sulphur in carbon disulphide in the presence of amyl alcohol, a crimson colour develops. *Cottonseed oil* and *kapok seed oil*, for example, give a positive reaction. Standard methods are described by AOCS [Cb 1 25(89)].

Halphenic acid

see *malvalic acid*

Hanus reagent

see *iodine number.*

hardening

a procedure to raise the melting point of oils and fats by partial hydrogenation. As a consequence of this some olefinic centres are hydrogenated and some *stereomutate* (change from *cis* to *trans* configuration) and move along the carbon chain. See *hydrogenation.*

hazelnut oil

an oleic-rich nut oil (ca 70%) from *Corylus avellana*. See also *speciality oils.*

HDL

see *lipoproteins.*

head space analysis

a method of chromatographic analysis of volatile material that collects in the gas phase when a liquid or solid sample is stored in a closed glass vessel. It is used as a measure of oxidation, for the detection of flavour compounds, and for the determination of residual solvent.

HEAR

high-erucic acid *rapeseed oil* typically containing ~50% of *erucic acid.*

Attempts are being made to produce oil with still higher levels (>65%) by *genetic modification*.

helenynolic acid

hydroxy acetylenic acid (9-OH-10*t*12*a*-18:2) present in *Helichrysum bracteatum* seeds.

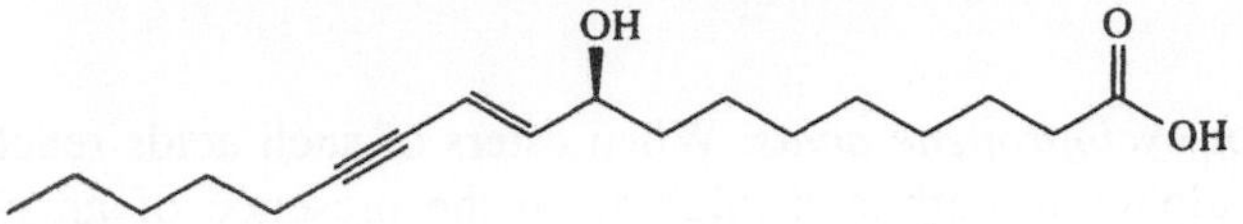

Helenynolic acid
9S-Hydroxy-10E-octadecen-12-ynoic acid
$C_{18}H_{30}O_3$
Mol. Wt.: 294,4

hepoxilins

epoxy hydroxy metabolites of 12-hydroperoxyeicosatetraenoic acid formed through the 12-lipoxygenase pathway.

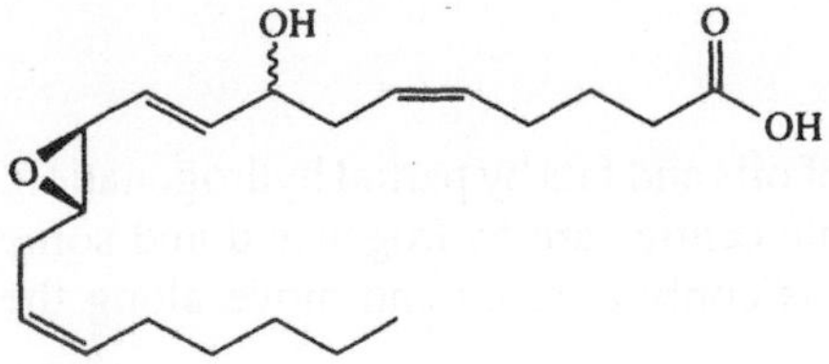

Hepoxilin A
8-Hydroxy-11S,12S-epoxy-5Z,9E,14Z-eicosatrienoic acid
$C_{20}H_{32}O_4$
Mol. Wt.: 336.5

heptadecanoic acid

the C_{17} acid (daturic, margaric, mp 61.3°C), present in many animal (especially ruminants) and vegetable oils but at quite low levels.

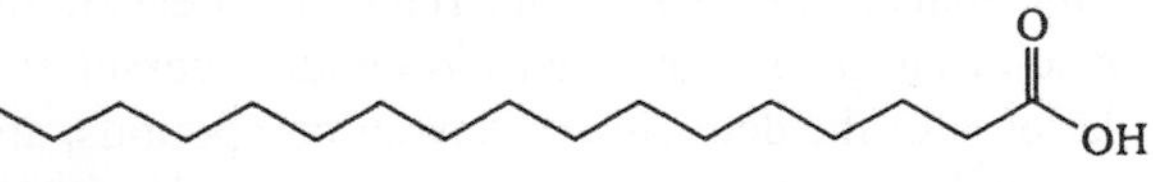

Margaric acid
heptadecanoic acid (17:0)
$C_{17}H_{34}O_2$
Mol. Wt.: 270.5

heptanal

the C_7 aldehyde produced along with *undecenoic acid* by pyrolysis of *ricinoleic acid*.

H-J. Caupin, Lipid Technologies and Applications (ed. F.D. Gunstone and F.B. Padley) Marcel Dekker, New York (1997), pp.787-795.

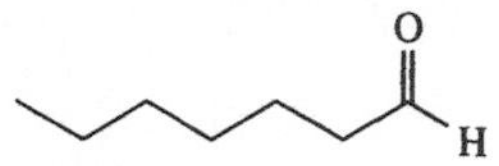

Heptanal
$C_7H_{14}O$
Mol. Wt.: 114.2

heptanoic acid

the C_7 acid (enanthic) produced by oxidation of *heptanal*.

H-J. Caupin, Lipid Technologies and Applications (ed. F.D. Gunstone and F.B. Padley) Marcel Dekker, New York (1997), pp.787-795.

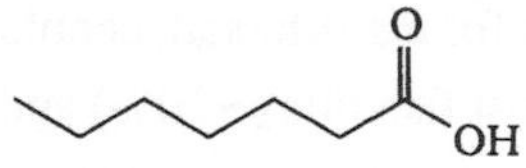

Enanthic acid
heptanoic acid
(7:0)
$C_7H_{14}O_2$
Mol. Wt.: 130.2

HERO

see *high-erucic rapeseed oil*.

herring oil

fish oil obtained from the herring, typically from the North Atlantic. See also *fish oils*.

HETE

see *hydroxyeicosatetraenoic acids*.

hexacosanoic acid

cerotic acid (26:0, mp 87–88°C).

hexadecanoic acid

the C_{16} acid (palmitic, mp 62.9°C) which is the most common of all saturated acids. It is present in virtually all animal and vegetable fats, especially in *palm oil* (~40%) and *cottonseed oil* (~25%). In vegetable oils this acid occurs in the *sn*–1 and *sn*–3 positions and hardly at all in the *sn*–2 position. *Lard* and most milk fats, including *human milk fat,* are unusual in that they have high levels of hexadecanoic acid in the *sn*–2 position (see *Betapol™*).

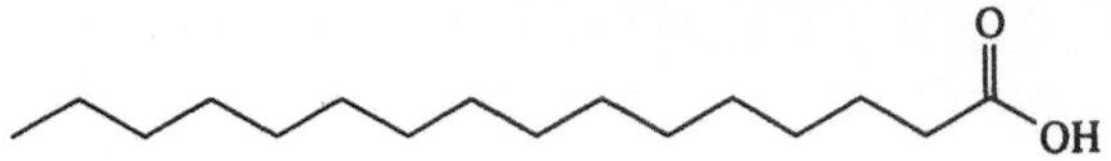

Palmitic acid
hexadecanoic acid
(16:0)
$C_{16}H_{32}O_2$
Mol. Wt.: 256.4

hexadecenoic acids
(i)   palmitoleic and zoomaric acids are trivial names for *cis*-9-hexadecenoic acid, the C_{16} analogue of *oleic acid*. It is present in most fish oils (~10%) and is a significant component of *macadamia oil* (~20%). Its *trans* isomer (palmitelaidic acid) melts at 32–33°C.

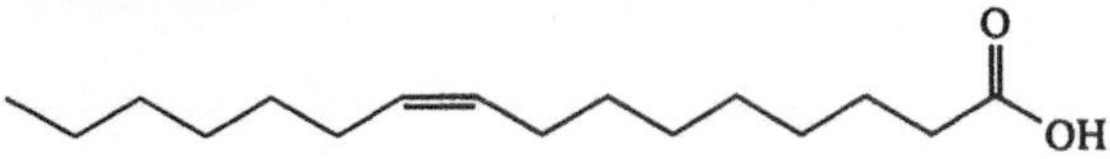

Palmitoleic acid, Zoomaric acid
9Z-Hexadecenoic acid
$C_{16}H_{30}O_2$
Mol. Wt.: 254.4

(ii) the *trans*-3 isomer (mp 53–54°C) occurs in leaf lipids in the phosphatidylglycerols at the *sn*–2 position..

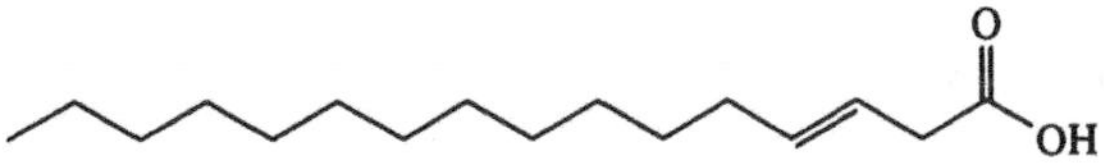

3*E*-Hexadecenoic acid
$C_{16}H_{30}O_2$
Mol. Wt.: 254.4

(iii) the Δ2 isomer, see *gaidic acid.*

hexagonal phases
liquid crystalline phases in polar lipid-water systems. Two phases exist, both with the lipids in cylindrical formations. H_I have the polar head groups on the surface and a continuous water phase. H_{II} have the inverted structure with the acyl chains as the continuous phase and water inside the cylinders.

K. Larsson, Lipids - Molecular Organisation, Physical Functions and Technical Applications, The Oily Press, Dundee, (1994).

H_I H_{II}

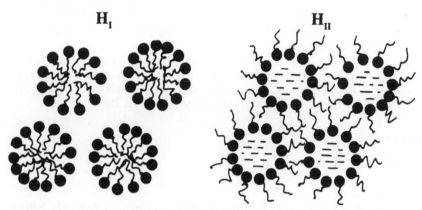

hexanoic acid
the C_6 acid (caproic) present at low levels in *lauric oils* and in *milk fat.*

Caproic acid
hexanoic acid
(6:0)
$C_6H_{12}O_2$
Mol. Wt.: 116.2

high-density lipoprotein
HDL, see *lipoproteins.*

high-erucic rapeseed oil
HERO, rapeseed oil with at least 50% *erucic acid.*

high-performance liquid chromatography
HPLC; chromatography with solvent mixtures as the mobile phase and

microparticles (3–10 μm) of, for example, silica or alkyl (most often C_{18}) bonded silica as the stationary phase. Normal (straight) phase HPLC refers to adsorption chromatography with columns packed with silica and non-polar solvent. Reversed-phase HPLC refers to columns with C_{18} material and polar solvent mixtures. This technique can also be used in silver ion mode. High pressure pumps are required for these columns. HPLC can be used for class separation of lipids (normal phase) as well as separation of molecular species of each class (reversed-phase). Several possibilities are available for the detection, e.g. UV-detectors, differential refractometers, *evaporative light-scattering detectors.*

P. Van der Meeren and J. Vanderdeelen, Advances in Lipid Methodology — Four (ed. W.W. Christie) The Oily Press, Dundee (1997) pp.83-118. B. Nikolova-Damyanova, Advances in Lipid Methodology — Four (ed. W. W. Christie) The Oily Press, Dundee (1997) pp.193-251. W.W. Christie, HPLC and Lipids — A Practical Guide, Pergamon Press, Oxford, 1987.

Highsun

see *Sunola.*

hiragonic acid
a trivial name given to hexadecatrienoic acids present in fish oils before detailed structures were assigned. It could be the 9,12,15-16:3 (*n*-1) or 6,9,12-16:3(*n*-4) isomer.

HLB
see *hydrophilic–lipophilic balance.*

^{1}H nuclear magnetic resonance spectroscopy
(i) using low resolution ^{1}H NMR spectroscopy it is possible to determine the *solid fat content* of a fat. Plotting solid fat content against temperature gives a *melting curve.* The system can also be modified to measure the oil content of seeds.

J. Warmsley, Simultaneous determination of oil and moisture in seeds by low-resolution pulsed NMR, Lipid Technlogy, 1998, 10, 135-137.

(ii) high resolution ^{1}H NMR spectra contain discrete signals for many of the hydrogen atoms (e.g. glycerol, olefinic, allylic, CH_2 α and β to the ester group). The spectrum can give information about the structure of the acids present in a natural mixture.

hormelic acid
see *cyclopentenyl acids.*

HpETE
see *hydroperoxyeicosatetraenoic acids.*

HPLC
see *high-performance liquid chromatography.*

Humicola
see *lipases.*

hydnocarpic acid
see cyclopentenyl acids.

hydration
the formal addition of water to an olefinic centre to give a hydroxy compound. This can be achieved by reaction with strong acid (protonation) followed by reaction with water. This is effected regiospecifically by a number of enzymes.

hydrocarbons
organic compounds containing only carbon and hydrogen. Included in this category are alkanes, alkenes and compounds such as *squalene* and *β-carotene.*

hydrogenation
the reaction of olefinic centres with hydrogen in the presence of nickel or other suitable metal catalyst. Partial hydrogenation is an important process for raising the melting points of liquid oils. At the same time, oxidative stability is enhanced by the reduction in the level of polyene acids but the nutritional value is reduced through reduction in *essential fatty acids* and increase in saturated and in *trans* acids. During partial hydrogenation some unsaturated centres are reduced whilst others undergo *stereomutation* and/or double bond migration. See also *hardening* and *biohydrogenation.*

W.T. Koetsier, Lipid Technologies and Applications (ed. F.D. Gunstone and F.B. Padley) Marcel Dekker, New York (1997), p 265-303. R.C. Hastert, Past, present and future of the hydrogenation process, Lipid Technology, 1998, 10, 101-105. D. Allen, Fat modification as a tool for product development. Hydrogenation and fractionation, Lipid Technology, 1998, 10, 29. M. Kellens, Edible Oil Processing (eds W. Hamm and R.J. Hamilton) Sheffield Academic Press, Sheffield (2000), pp.129-143.

hydrogenolysis
a process by which long-chain acids or esters are reduced to *alcohols* required

as *surfactants*. The reaction occurs at high temperature and pressure (250–300°C and 200–300 bar) in the presence of copper or zinc chromite as catalyst. The reaction may be accompanied by *hydrogenation*, i.e. reduction of olefinic groups.

Fatty Alcohols, Raw materials, Uses; Henkel KgaA, Dusseldorf, Germany, 1981.

hydrolysis
the splitting of esters (and amides) with water, usually in the presence of an acidic or basic catalyst. The weight relationship between triacylglycerol and derived fatty acid is given typically by: hydrolysis of triolein (100 g) requires water (6.1 g) to give oleic acid (95.7 g) and glycerol (10.4 g). See also *lipolysis, saponification, fat splitting.*

hydroperoxides
these are the first products formed during autoxidation of olefinic acids. They are allylic in nature and the double bond(s) may have changed configuration and/or position compared with the original olefinic acid. These compounds can be detected and measured as the *peroxide value*. They do not themselves have undesirable flavour or odour but are readily degraded to short-chain compounds with these properties.

E.N. Frankel, Lipid Oxidation, The Oily Press, Dundee (1998).

Hydroperoxides

hydroperoxy acids
acids containing a *hydroperoxide* group (OOH).

hydroperoxyeicosatetraenoic acids (HpETE)
hydroperoxides produced from arachidonic (*eicosatetraenoic*) acid. Six isomers are known with the *hydroperoxy* group at C-5, 8, 9, 11, 12 or 15 and appropriate tetraene systems containing one conjugated diene unit, e.g. 5-OOH 6, 8,11,14-eicosatetraenoic acid.

hydrophilic
literally water-loving. Molecules or parts of molecules which are soluble in water. In *amphiphilic* compounds the hydrophilic portion is described as a head group and may be anionic, non-ionic, cationic or amphoteric.

hydrophilic–lipophilic balance

HLB; measures the size and strength of the hydrophilic and lipophilic parts of an emulsifier, determined by the chemical composition and ionization strength of the emulsifier:

HLB = 20[mol weight of the hydrophilic portion/mol weight of the molecule]

or HLB = 20(1–S/A) where S is the *saponification value* and A is the acid number of the fatty acid portion of the emulsifier. Typically, emulsifiers (5–6), wetting agents (7–9), and detergents (13–15) have the range of HLB values shown.

G. Bognolo, Lipid Technologies and Applications (ed. F.D. Gunstone and F.B. Padley) Marcel Dekker, New York (1997), p.670.

hydrophilisation

an industrial process for obtaining concentrates of *oleic acid* from the fatty acids of *tallow* or *palm oil.* After crystallization at about 20°C a wetting agent is added. The aqueous suspension containing the crystallised solids is separated from the liquid fraction by centrifugation. The liquid fraction is mainly *oleic acid* (70–75%) with hexadecenoic and linoleic acids and less than 10% of saturated acids.

hydrophobic

literally water-hating. Molecules or parts of molecules which are insoluble in water and are usually lipophilic. They are generally long alkyl chains derived from lipids or made by olefin oligomerisation.

hydroxy acids

(i) acids containing a hydroxyl group. *Ricinoleic acid* is the best known member of this class but others include *aleuritic, cerebronic, coriolic, densipolic, dimorphecolic, kamlolenic, lesquerolic acid* and *many eicosanoids and lipoxins.*

$$CH_3(CH_2)_m \quad \overset{OH}{\diagup} \quad (CH_2)_n \quad \overset{O}{\diagup} OH$$

Hydroxy acids

(ii) ω-hydroxy acids (C_{14}–C_{22}) occur in waxes. See, for example, *phellonic* and *juniperic*. They can be converted to macrocyclic lactones.

(iii) hydroxyeicosatetraenoic acids obtained by reduction of the *hydroperoxyeicosatetraenoic* acids. The three major acids of this type (5-, 12-, and 15-HETE) result from the action of the 5-, 12-, and 15-*lipoxygenase* enzymes.

(iv) hydroxyheptadecatrienoic acid (12-OH 5,8,10-17:3) is a metabolite of prostaglandin H$_2$. It is readily oxidised to the corresponding oxo acid.

(v) hydroxyoctadecanoic acid can exist in several isomeric forms but the best known is 12-hydroxystearic acid produced by hydrogenation of *ricinoleic acid*. It is used in greases as the lithium or other salt. See also *rosilic acid.*

hydroxylation

(i) the chemical conversion of alkenes to *vic*-dihydroxy compounds. Reagents such as osmium tetroxide or potassium permanganate effect *cis* addition. *Trans* hydroxylation is best effected by *epoxidation* followed by hydrolysis.

Hydroxylation

After further derivatisation the diols furnish compounds useful for mass spectrometric examination with consequent identification of double bond position, though this procedure is now seldom used for this purpose.

(ii) the biological introduction of hydroxyl groups into polyunsaturated acids through *lipoxygenase*-promoted oxidation. Hydroxylation should not be confused with *hydration* which is another route to hydroxy acids.

hydroxyl value
the number of mg of potassium hydroxide required to neutralise the acetic acid capable of combining by acetylation with one gram of oil or fat.

hypercholesterolaemia
a disease in which the plasma cholesterol concentration increases characteristically due to a slow rate of breakdown of the low-density lipoproteins (LDL particles). This condition is a risk factor for coronary heart disease.

hyperlipidaemia
general term for metabolic defects leading to elevated plasma concentrations of lipoproteins (hyperlipoproteinaemias), resulting in high concentration of triacylglycerols (hypertriacylglycerolaemia) or cholesterol (hypercholesterolaemia) or both. These conditions are risk factors for cardiovascular disease.

G.R. Thompson, A Handbook of Hyperlipidaemia: Detection of Hyperlipidaemia and Benefits of Treatment, (ed. G.R. Thompson), Current Science, (1989), pp.209-223. P. Layer, Pancreatic enzymes: secretion and luminal nutrient digestion in health and disease, Journal of Clinical Gastroenterology, 1999, 28, 3-10. G.J. Miller, Hyperlipidaemia and hypercoagulability, Progress in Lipid Research, 1993, 32, 61-69.

I

Iatroscan
commercial name (Iatron, Japan) for an instrument incorporating a flame-ionization detector and utilizing quartz rods coated with a layer of fused silica or alumina as the chromatographic medium. The separation is performed on the rods in a similar manner to *thin layer chromatography* on plates and the lipids are detected by subsequent combustion in the detector flame.

icosanoic acid
eicosanoic acid. See *arachidic acid.*

icosanoids
see *eicosanoids.*

icosapentaenoic acid
see *eicosapentaenoic acid.*

icosatetraenoic acid
see *eicosatetraenoic acid.*

icosatrienoic acid
see *eicosatrienoic acid.*

icosenoic acid
see *eicosenoic acid.*

Illipé butter
fat from the seeds of *Shorea stenoptera.* It resembles *cocoa butter* in its fatty acid composition and melting behaviour. See also *Shorea robusta.*

INFORM
International News on Fats, Oils, and Related Material. Published by the *American Oil Chemists' Society* since 1990.

infrared spectroscopy
spectra covering the range 2.5–15m^{-6}. Most used with lipids to detect and measure *trans unsaturation* (frequency 968 cm^{-1}, wavelength ~10.3 x 10^{6}m).

INRA
Institut National de la recherche Agronomique in France.

interchangeability
the idea that many oils are similar enough in composition and properties, or can be made so by suitable modification, to be used in place of one another. This allows use of more readily-available and/or cheaper material in place of less readily available and/or more expensive products. In considering interchangeability attention must be given to (a) chemical composition in terms of fatty acids and triacylglycerols, (b) physical properties — especially melting behaviour, content of solids, and crystal form — and (c) nutritional properties — especially levels of saturated, monoene, *trans*, polyene acids, and of minor components also present.
K. Berger Functionality and interchangeability of fats,, Lipid Technology, 1989, 1, 40-43.

interesterification
term given to the production of esters by interaction of two esters in the presence of an alkaline or enzymic catalyst. The alkaline catalyst is usually an alkali metal (0.2–0.3%) and reaction is generally effected at 80–90°C over half an hour. Using oils and fats (i.e. triacylglycerol mixtures) interesterification leads to *randomisation* of the acyl groups. Using a 1,3-stereospecific enzyme, change is confined to the *sn*–1 and *sn*–3 positions. Valuable applications of this specificity include *Betapol*^{TM}, products with improved melting behaviour, and incorporation of medium chain length acids and eicosapentaenoic acid into a vegetable oil not containing these acids).
A. Rozendaal and A.R. Macrae, Lipid Technologies and Applications (ed. F.D. Gunstone and F.B. Padley) Marcel Dekker, New York (1997), pp.223-263. D.A. Allen, Interesterification — a vital tool for the future, Lipid Technology, 1996, 8, 11-15; Fat modification as a tool for product development. Interesterification and biomodification, Lipid Technology, 1998, 10, 53-57. D. Rousseau and A.G. Maringoni, Food Lipids: Chemistry, Nutrition, and Biotechnology, (ed. C.C. Akoh and D.B. Min) Marcel Dekker, New York (1998) pp.251-281. W.M. Willis and A.G. Marangoni, Food Lipids: Chemistry, Nutrition, and Biotechnology, (ed. C. C. Akoh and D.B. Min) Marcel Dekker, New York (1998) pp.665-698. G.P. Mc Neill, Lipid Synthesis and Manufacture (ed. F.D. Gunstone) Sheffield Academic Press, Sheffield (1999) pp.288-320. M. Kellens, Edible Oil Processing (eds W. Hamm and R.J. Hamilton) Sheffield Academic Press, Sheffield (2000), pp.144-155.

International Society for Fat Research.
ISF, started by H.P. Kaufmann in Germany, and first largely confined to Europe, it is now an international organisation with a secretariat provided by

the *American Oil Chemists' Society.* It organises international conferences about every two years.

International Standards Organisation
an organisation based on over 70 national organisations for standards. It has sub-committees covering oil seeds, fats, soaps etc.

Intralipid™
commercial name for a fat emulsion for clinical use. See *intravenous fat emulsions.*
R.T. Lyons and E.G. Carter, Lipid Technologies and Applications (ed. F.D. Gunstone and F.B. Padley) Marcel Dekker, New York (1997), pp.535-556.

intravenous fat emulsions
fat emulsions made of soybean oil (*Intralipid*), safflower oil (Liposyn) and phospholipids from egg yolk or soybean lecithin.
R.T. Lyons and E.G. Carter, Lipid Technologies and Applications (ed. F.D. Gunstone and F.B. Padley) Marcel Dekker, New York (1997), pp.535-536.

iodine number (value)
classical method to obtain information on the degree of unsaturation in fats and oils. An iodine-bromine (*Hanus reagent*) or iodine monochloride (*Wijs reagent*) reagent is reacted with the double bonds and the excess reagent (as iodine) is titrated with sodium thiosulphate solution. Standard methods are described by AOCS [Cd 1 25(89) and Cd 1b 87(89)] and IUPAC (2.205). See also *Fourier-Transform Infrared Spectroscopy.*

ipurolic acid
trivial name for 3,11-dihydroxytetradecanoic acid. This acid is present in oils from *Ipomoea* species.

irradiation
exposure of foods to ionizing radiation so that microorganisms and pests are controlled with minimum change in the food.

isanic acid
trivial name for 9a11a17e-18:3. Present in *Onguekoa gore* seed oil. Also called erythrogenic acid.

isanolic acid
the 8-hydroxy derivative of *isanic acid.*

ISF
see *International Society for Fat Research.*

ISO
see *International Standards Organisation*

iso acids
(i) in terms of strict organic nomenclature iso acids have a branched methyl group at the ω-2 carbon atom. The natural members generally have an even number of carbon atoms in each molecule and are biosynthesised by chain-elongation of the C_4 acid 2-methylpropanoic (isobutyric), itself a protein metabolite.

Iso acids

(ii) the term iso is used more loosely to indicate a compound of isomeric nature: e.g. *iso-oleic acid* is a mixture of 18:1 acids produced during partial hydrogenation of unsaturated C18 acids and *isostearic acid* is a byproduct of the production of *dimer acids.*

isoprene unit
2-methylbutadiene; the branched-chain C_5 structural unit of terpenoid compounds including sterols (isoprenoids).

Isoprene
2-methylbutadien
C_5H_8
Mol. Wt.: 68.1

isovaleric acid
a C_5 branched-chain acid (3-methylbutanoic) present in the lipids of the melon of porpoises and dolphins.

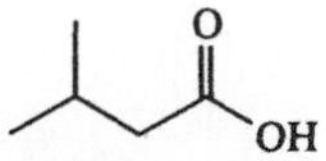

Isovaleric acid
3-Methylbutanoic acid
$C_5H_{10}O_2$
Mol. Wt.: 102.1

ITERG

Institute des Corps Gras. A French group concentrating on the scientific and technical aspects of lipids. Was the publisher of *Revue Française des Corps Gras* which was replaced in 1994 by *Oleagineux Corps gras Lipides*.

J

jacaric acid
see *conjugated unsaturation.*

jalapinolic acid
trivial name for 11-hydroxyhexadecanoic acid [mp 68–69°C(S) 68–69°C(RS)], also called scammonolic acid and turpetholic acid. The S form occurs in jalap and scammony resins and also as a range of sugar derivatives (muracatins and operculinic acid).

JAOCS
see *Journal of the American Oil Chemists' Society.*

jasmonic acid
a C_{12} cyclopentane acid which serves as a plant hormone. A metabolite of linoleic acid.

A. Grechkin, *Recent developments in biochemistry of the plant lipoxygenase pathway, Prog. Lipid Research, 1998, 37, 317-352.*

Jasmonic acid
3-Oxo-2R-(2Z-pentenyl)-1R-cyclopentaneacetic acid
$C_{12}H_{18}O_3$
Mol. Wt.: 210.3

jojoba oil
the seed oil of *Simmondsia chinensis* which grows in desert regions under arid conditions. The seed oil is not a triacylglycerol but a *wax ester* made up mainly of 18:1 (6%), 20:1 (35%) and 22:1 (7%) acids with 20:1 (22%), 22:1 (21%) and 24:1 (4%) alcohols. It is thus a mixture of C_{38}–C_{44} esters with one double bond in each part of the molecule. The absence of methylene-interrupted polyene material makes the oil oxidatively stable. It is used in cosmetics and has excellent lubricating properties.

Journal of Lipid Mediators
journal published by Elsevier Science Publishers (Amsterdam) since 1989.

Journal of Lipid Research
a monthly journal published by Lipid Research Inc (Rockville Pike, Bethesda, Maryland, USA) since 1959.

Journal of Liposome Research
journal published by Marcel Dekker, New York, Inc, (New York, NY, USA) since 1989.

Journal of Surfactants and Detergents
journal published by the *American Oil Chemists' Society* starting in 1998.

Journal of the American Oil Chemists' Society
one of four publications of the *American Oil Chemists' Society* (publishes also *INFORM, Lipids* and *Journal of Surfactants and Detergents*). Formerly known as Oil and Soap (1932–47) and Journal of the Oil and Fat Industries (1924–31).

juniperic acid and juniperinic acid
trivial names for 16-hydroxyhexadecanoic acid present in conifer waxes. Its lactone is dihydroambrettolide. See also *hydroxy acids*.

juniperonic acid
trivial name of 5c11c14c17c-20:4 (5,11,14,17-*eicosatetraenoic* acid).

K

kamala oil
seed oil from *Mallotus phillipinensis* containing *kamlolenic acid.*

kamlolenic acid
trivial name for the hydroxy fatty acid 18-OH 9*c*11*t*13*t*-octadecatrienoic acid (mp 78°C, all-*trans* isomer mp 89°C) present in kamala oil.

kapok seed oil
seed oil from the tropical trees *Ceiba pentandra (Eriodendron anfractuosum)* and *Bombax malarbaricum* , both of the family Bombacaceae. The seeds are a by-product of kapok fibre production. The oil contains *cyclopropene acids.*

karite
see *Butyrospermum parkii.*

kephalin
cephalin; old term for *phosphatidylethanolamine.*

kokum butter
produced from the seeds of the Kokum tree (*Garcinia indica choisy*) grown in India. The fat is used as a *cocoa butter* substitute. See *chocolate* and *cocoa butter alternatives.*

Krabbe's disease
metabolic disorder resulting in the accumulation of galacto*cerebrosides*. See also *glycosylceramides.*

Krafft point
the lowest temperature before crystallisation of a micellar solution.

Kreis test
an older method of detecting oxidation now not much used. When oxidized material is shaken with a dilute ether solution of phloroglucinol (1,3,5-trihydroxybenzene) in the presence of hydrochloric acid a pink colour develops. The mechanism of this change has not been identified.

L

lacceric acid
trivial name for dotriacontanoic acid (32:0). Present in wool fat and in several leaf waxes.

lacceroic acid
another trivial name for dotriacontanoic acid (32:0). See *lacceric acid.*

lactarinaric
the trivial name for 6-oxo-octadecanoic acid (mp 87°C) present in the lipids of *Lactarius rufus.*

Lactem™
made by reaction of mono/diacylglycerols or distilled monoacylglycerol with 15–35% lactic acid. A simplified structure is shown below but compounds with 2–6 lactic acid units will also be present. Used mainly in speciality fats for dessert products such as non-dairy creams, toppings, cake-mixes etc. E number 472(e) in Europe.
N. Krog, Lipid Technologies and Applications (ed. F.D. Gunstone and F.B. Padley) Marcel Dekker, New York (1997), pp.521-534.

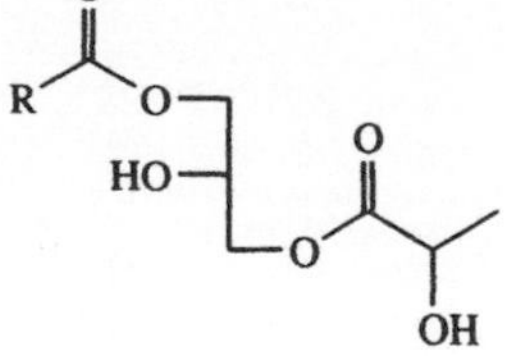

Lactem™

lactobacillic acid
phytomonic acid; 11,12-methyleneoctadecanoic acid. This cyclopropane acid occurs in the *cis* form in certain microorganisms, mainly gram-positive bacteria. It is derived biochemically from 11*c-octadecenoic acid* (*cis*-vaccenic acid). (See structure on the next page.)

Lactobacillic acid
2-Hexylcyclopropanedecanoic acid
$C_{19}H_{36}O_2$
Mol. Wt.: 296.5

lamellar phase
liquid-crystalline phase in polar lipid-water systems. Lipid bilayers alternate
with water in a lamellar structure.
*K. Larsson, Lipids – Molecular Organisation, Physical Functions and Technical Applications,
The Oily Press, Dundee, (1994).*

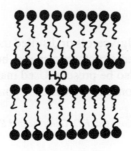

Lamellar phase

lampante
see *olive oil (grades).*

Langmuir-Blodgett films
monolayer of lipids on a water surface, which may be created and studied in
a surface balance, a trough with water and a moving barrier to compress the
film. During compression a monolayer undergoes transitions which are called
gaseous phase, liquid phase and solid phase.
*K. Larsson, Lipids — Molecular Organisation, Physical Functions and Technical Applications,
The Oily Press, Dundee (1994).*

lanolin
see *woolwax.*

lanosterol

a C_{30} sterol. In the biosynthesis of cholesterol from mevalonate via *squalene*, lanosterol is the first tetracyclic compound in this pathway.

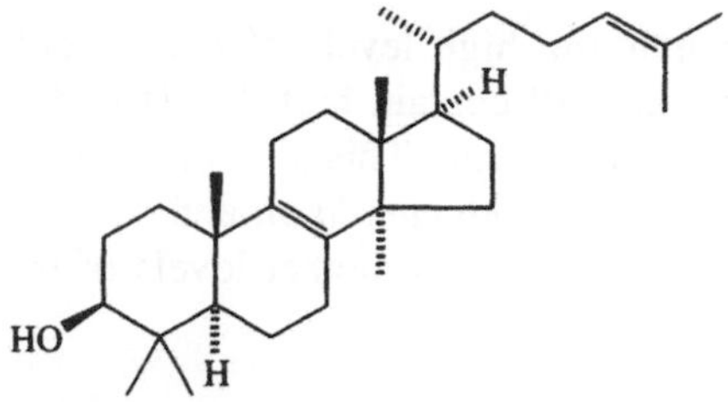

Lanosterol
Lanosta-8,24-dien-3β-ol
$C_{30}H_{50}O$
Mol. Wt.: 426.7

Lanza fractionation

the separation of *stearin* and *olein* resulting from dry *fractionation* is facilitated by the addition of an aqueous solution of detergent. The solid stearin, coated by detergent, goes into the aqueous phase. This procedure is no longer much used though in a modified form (*hydrophilisation*) it is applied to the separation of solid and liquid acids.

lard

a solid fat obtained from the depot fat of pigs. Available in quantities of about 6 million tonnes a year. The fat contains palmitic (20–32), stearic (5–24), oleic (35–62), and linoleic acid (3–16%) as major components. It is unusual in that 70% of the palmitic acid is in the *sn*–2 position. It also contains *cholesterol* (0.37–0.42%).

large unilamellar vesicles

LUV, see *liposomes*.

La Rivista Italiana delle Sostanze Grasse

journal published by the Italian Society for Lipid Research (Società Italiana per lo Studio delle Sostanze Grasse).

lauric acid
trivial name for *dodecanoic acid* (12:0, mp 44.8°C). Found mainly in *lauric oils* such as *coconut* and *palm kernel*.

lauric-canola
a genetically modified *rapeseed oil* containing high levels of lauric acid. Commercial crops from early development still contain high levels of C_{18} acids (ca 55%) along with 40–45% of C_{12} and C_{14} acids. This oil is different in fatty acid and triacylglycerol composition from both *canola oil* and *coconut oil*. It is expected that crops developed later will have higher levels of the medium-chain acids.

lauric oils
oils rich in lauric acid, such as *coconut* and *palm kernel*. They are used in food products and in the manufacture of soaps and detergents. See also *lauric-canola*.

lauroleic acid
trivial name for *cis*-9-dodecenoic acid (12:1).

LCAT
see lecithin cholesterol acyltransferase

LDL
low-density lipoproteins. See *lipoproteins*.

lecithin
refers to a crude industrial mixture of phospholipids obtained from vegetable oils and from egg yolk (the name *lecithin* comes from the Greek word for egg yolk — *lekithas*). Most lecithin is derived from soybean oil and is recovered during the *degumming* process. Typically it is a mixture of phospholipids (47%), triacylglycerols (36%), other lipids (11%), carbohydrates (5%) and water (1%). The phospholipids are mainly *phosphatidylcholines*, *phosphatidylethanolamines* and *phosphatidylinositides*. The level of phospholipids can be raised by *de-oiling*. Lecithins are designated E number 322.

M. Schneider, Industrial production of phospholipids-lecithin processing, Lipid Technology, 1997,9, 109-116. M. Schneider, Lipid Technologies and Applications (ed. F.D. Gunstone and F.B. Padley) Marcel Dekker, New York (1997), pp.51-78. E.F. Sipos and B.F. Szuhaj, Bailey's Industrial Oil and Fat Products, (ed. Y.H. Hui) John Wiley & Sons, New York (1996), Volume 1, pp.311-395.

lecithin cholesterol acyltransferase (LCAT)
the function of this enzyme is to transfer a fatty acid from phosphatidylcholine to cholesterol in the *HDL lipoprotein* particle.

Lesquerella oils

contain *lesquerolic acid* (>50%) with lower levels of other hydroxy acids. Attempts are being made to develop *Lesquerella fendleri* as a commercial crop.

lesquerolic acid
the C_{20} homologue of *ricinoleic acid* (*R cis* isomer) present in seed oils of the *Lesquerella* species.

Lesquerolic acid
14*R*-Hydroxy-11Z-eicosenoic acid
$C_{20}H_{38}O_3$
Mol. Wt.: 326.5

leukotoxins
epoxyoctadecenoic acids derived from linoleic acid and observed, for example, in samples of burned skin. Leukotoxin A is 9,10-epoxy-12-octadecenoic acid (*coronaric acid*) and leukotoxin B is 12,13-epoxy-9-octadecenoic acid (*vernolic acid*).

leukotrienes
acyclic metabolites of arachidonic acid (*eicosatetraenoic acid*). Leukotriene A₄ (an epoxide) can be hydrolysed to leukotriene B₄ or converted to the tripeptide (leukotriene C₄). In turn this can be converted to a dipeptide (leukotriene D4) and a monopeptide (leukotriene E₄).
N.H. Wilson, Lipid Synthesis and Manufacture (ed. F.D. Gunstone) Sheffield Academic Press, Sheffield (1999) pp.127-161.

Leukotriene A₄
5S,6S-Epoxy-7E,9E,11Z,14Z-eicosatetraenoic acid
$C_{20}H_{30}O_3$
Mol. Wt.: 318.5

licanic acid

a C_{18} keto acid with triene conjugation, 4-oxo-9c11t13t-*octadecatrienoic acid*. Present in oiticica oil *Licania rigida*.

light-scattering detectors

see *evaporative light scattering detector*.

lignoceric acid

trivial name for *tetracosanoic acid* (24:0 mp, 84.2°C). It occurs in some waxes and is a minor component of some seed oils.

Limnanthes alba

see *meadowfoam oil*.

lin (huile de)

French name for *linseed* oil.

linelaidic acid

trivial name for 9t12t-*octadecadienoic acid* (mp 28–29°C). It occurs naturally only very rarely (e.g. *Chilosis linearis* seed oil) but can be made from linoleic acid by *stereomutation*.

linderic acid

a C_{12} monoene acid, 4c-12:1, present in *Lindera obtusiloba* seed oil.

Linola™ oil

oil from a modified flaxseed. It differs from the more common *linseed oil* in having a high level of linoleic acid and a low level of α-linolenic acid. The generic name for this flaxseed is *solin*. It was developed in Australia and is now grown in Canada and Europe.

A.G. Green and J.C.P. Dribnenki, Linola — a new premium polyunsaturated oil, Lipid Technology, 1994, 6, 29-33.

linoleic acid

trivial name for 9c12c-*octadecadienoic acid*. See *octadecadienoic acids*.

linolenic acid

this *octadecatrienoic acid* exists in two isomeric methylene-interrupted forms: α-linolenic acid (n–3) and γ-linolenic acid (n–6). See *octadecatrienoic acids*.

linseed oil
the oil obtained from *Linus usitatissimum*. This valuable drying oil (~0.7 million tonnes a year) is produced in Argentina, India, USSR, USA, and Canada with Argentina and Canada being the most important exporting countries. The oil has a high *iodine value* (~177) and contains 55–60% of α-*linolenic acid*. As a consequence of this, the major triacylglycerols are typically LnLnLn (35), LnLnL (14), LnLnO (19), and LnLnS (7%). Linseed oil is used in paints and varnishes, in the production of linoleum, and as a sealant for concrete. It is available as a food oil under the name of flax seed oil. A modified form of the seed (*solin*) gives an oil with a high level of linoleic acid and hardly any α-linolenic acid (*Linola oil*). This is used in the food industry as an alternative to other linoleic-rich oils.

B.E. Prentice and M.D. Hildebrand, Exciting prospects for flax and linseed oil, Lipid Technology, 1991, 3, 83-89.

Linum usitatissimum (Linaceae)
this is flax which exists in two strains. One is grown for the fibre (flax) and the other for its seed oil (*linseed oil*) and valuable protein.

lipases
enzymes (catalysts) which promote the *hydrolysis* of lipids. Under experimental conditions they also catalyse *acidolysis, alcoholysis* or *interesterification* of lipids. They are often specific for particular ester bonds and, under appropriate reaction conditions, can catalyse hydrolysis or esterification. Lipases are becoming more important because of the specificity of their reactions and because reaction products are relatively free of chemical reagents. They are increasingly available in immobilised form on a commercial scale. Examples of useful enzymes are the lipases from: *Candida cylindricae, Corynebacterium acnes* and *Staphylococcus aureus* which are all non-specific; *Aspergillus niger, Mucor javanicus, M. miehei, Rhizopus arrhizus, R. delamar* and *R. niveus* all of which are 1,3-regiospecific; and *Geotrichium candidum* which is Δ9-specific. Many of these enzymes discriminate against acids with unsaturation close to the carboxyl group (e.g. Δ4 acids such as docosahexaenoic, Δ5 acids such as eicosapentaenoic and arachidonic, and Δ6 acids such as γ-linolenic and stearidonic) and can be used to enhance the level of these acids in mixtures in which they are present as acids or esters. See also *phospholipases* and *enzymic enhancement*.

T. Godfrey, Lipases for industrial use, Lipid Technology, 1995, 7, 58-61. K.D.Mukherjee, Food Lipids: Chemistry, Nutrition, and Biotechnology, (ed. C.C. Akoh and D.B. Min) Marcel Dekker, New York (1998) pp.589-640. J. D. Weete, Food Lipids: Chemistry, Nutrition, and Biotechnology, (ed. C.C. Akoh and D.B. Min) Marcel Dekker, New York (1998) pp.641-664. G.P. McNeill, Lipid

Synthesis and Manufacture (ed. F.D. Gunstone) Sheffield Academic Press, Sheffield (1999) pp.288-320. K.D. Mukherjee, Plant lipases and their application in lipid biotransformations, Progress in Lipid Research, 1994, 33, 1/2, 165-174. F.D. Gunstone, Enzymes as biocatalysts in the modification of natural lipids, J. Sci. Fd. Agric., 1999, 79, 1535-1549.

lipid A
a polymeric glycolipid present in the membranes of gram-negative bacteria.

lipid bodies
see *oil bodies*.

lipid class
the broad group of *lipids* is subdivided into classes depending on chemical structure, *e.g. diacylglycerols, phosphatidylethanolamines, ceramides* etc. They can be separated into individual molecular species based on their acyl groups.

lipid digestion
the process to break down dietary lipids into their component parts by enzymes. Triacylglycerols, the major part of the dietary lipids, are efficiently hydrolysed in the small intestine by pancreatic lipase and the phospholipids by phospholipases. It is generally accepted that monoacylglycerols, monoacylphospholipids, and fatty acids together with bile salts form mixed micelles which are absorbed by the enterocytes in the jejunum.

J.F. Mead, Lipids. Chemistry, Biochemistry and Nutrition: Digestion and Adsorption of Lipids, (eds) Plenum Press (1986), pp.255-272.

LipidForum
The Scandinavian society for lipid research and technology founded in 1952.

lipidoses
rare diseases in which lipids, normally glycosphingolipids, accumulate in tissues because of the malfunctioning of specific catabolic enzymes.

Lipids
the title of one of four journals published by the *American Oil Chemists' Society*. First published in 1966.

lipids
the common name for a broad group of natural products. They are mainly *hydrophobic* though some are *amphiphilic*. (See Preface).

W.W Christie, <www.lipid.co.uk/INFORES/index.html>

Lipid Technology
journal published by P.J. Barnes and Associates (Bridgwater, England) since 1989 along with *Lipid Technology Newsletter* since 1995.

lipoamino acids
a group of lipophilic compounds incorporating α-amino acids. Examples include aminoacyl esters of phosphatidylglycerol and acylornithines (ornithine amide lipids), present in gram-positive bacteria.

Typical acylornithine

lipofrac process
see *Lanza fractionation*.

lipolysis
the complete or partial hydrolysis of lipids catalysed by *lipases*.

lipophilic
molecules or parts of molecules which are soluble in fatty material and usually insoluble in water. See also *hydrophilic* and *hydrophobic*.

lipopolysaccharides
polymeric structure consisting of sugar units linked to *lipid A* in Gram-negative bacteria.

S.G. Wilkingson, Bacterial Lipopolysaccharides — Themes and Variations, Progress in Lipid Research, 1996, 35, 283-343.

lipoprotein lipase
a lipase which specifically acts on the triacylglycerols of *lipoproteins*.

lipoproteins

lipoproteins are complexes of lipids with specific proteins (*apoproteins*) whose function is to transport lipids between organs in an aqueous environment. There are four main classes of lipoprotein particles present in human plasma. These are defined according to their flotation properties on ultracentrifugation in media of specific density: (i) chylomicrons (density <0.95) are produced in the intestine from dietary lipids and are rich in triacylglycerols (~83%), (ii) very-low-density lipoproteins (VLDL, density range 0.95–1.006) are synthesised in the liver and contain triacylglycerols (~50%), phospholipids (~20%) and cholesterol (~22%), (iii) low-density lipoproteins (LDL, density range 1.019–1.063), containing triacylglycerols (~10%), phospholipids (~22%) and cholesterol (~48%), carry cholesterol to peripheral tissues, and (iv) high-density lipoproteins (HDL, density range 1.063–1.210), with triacylglycerols (~8%), phospholipids (~22%) and cholesterol (~20%), carry cholesterol back to the liver.

M.D. Feher, W. Richmond, Pocket Picture Guides. Lipids and Lipid Disorders: Lipoproteins: structure and function, Gower Medical Publishing, Aldershot (1990), pp.5-12. M.D. Feher, W. Richmond, Pocket Picture Guides. Lipids and Lipid Disorders: Lipids, lipoproteins and vascular disease, Gower Medical Publishing, Aldershot (1990), pp.44-58. G.R. Thompson, A Handbook of Hyperlipidaemia, Plasma lipids and lipoproteins, Current Science, London (1989), pp.3-21. P.J. Barter, New Trends in Lipid and Lipoprotein Analysis (eds J.-L. Sebedio and E.G. Perkins), AOCS Press, Champaign, Illinois (1995), pp.337-345. J. Babiak, L.L. Rudel, Baillère's Clinical Endocrinology and Metabolism. International Practice and Research. Vol. 1, No. 3: Lipoprotein Metabolism, (ed. J. Shepherd), Baillère Tindall, (1987), pp.515-550. G.V. Marinetti, Disorders of Lipid Metabolism. Cholesterol, Lipoproteins, and Atherosclerosis, Plenum Press (1990), pp.121-134.

liposomes

spherical particles formed from lamellar lipid-water systems. They consist of closed *bilayers* with water inside. One bilayer (small and large unilamellar vesicles, SUV and LUV) or several bilayer (multilamellar vesicles, MLV) particles can exist. Use of liposomes in drug delivery and targeting is being investigated extensively.

J.R. Philippot et al., Liposomes as Tools in Basic Research and Industry: Liposomes as cell or membrane models, (eds J.R. Philippot and F. Schuber) CRC Press, Boca Raton, Florida (1995), pp.71-156. J.R. Philippot et al., Liposomes as Tools in Basic Research and Industry: The Liposomes, (eds J.R. Philippot and F. Schuber) CRC Press, Boca Raton, Florida (1995), pp.3-67. J.R. Philippot et al., Liposomes as Tools in Basic Research and Industry: New developments of liposomes, (eds J.R. Philippot and F. Schuber) CRC Press, Boca Raton, Florida (1995), pp.159-263. S. Hirota, New Liposome Developments, Advanced Drug Delivery Reviews, 1997, 24, 2-3, 123-363.

lipoteichoic acids
glycolipids linked to teichoic acid (a polymer of glucose-1-phosphate), present in bacterial surface membranes.

lipoxins
trihydroxyeicosatetraenoic acids produced from arachidonic acid (*eicosatetraenoic acid*) by enzymic oxidation, epoxide formation, and subsequent hydrolysis. The lipoxins contain a conjugated tetraene system. Examples are lipoxin A_4 (LXA$_4$; 5,6,15-trihydroxy-7,9,11,13-20:4) and lipoxin B_4 (LXB$_4$; 5,15,16-trihydroxy-6,8,10,12-eicosatetraenoic acid).

N.H. Wilson, Lipid Synthesis and Manufacture (ed. F.D. Gunstone) Sheffield Academic Press, Sheffield (1999), pp.127-161.

Lipoxin A$_4$
5S,6R,15S-Trihydroxy-7E,9E,11Z,13E-eicosatetraenoic acid
$C_{20}H_{32}O_5$
Mol. Wt.: 352.5

lipoxygenase
iron-containing enzymes which promote the oxidation of polyenoic acids to *hydroperoxides* usually in a regiospecific and stereospecific manner. Lipoxygenases from several plant sources (soybean, corn, potato, tomato, etc) convert *linoleic* acid to its 9- and/or 13-hydroperoxide. Other lipoxygenases, operating particularly with arachidonic acid (*eicosatetraenoic acid*), produce hydroperoxides of this acid which easily undergo further reaction. See *hydroperoxyeicosatetraenoic acids*.

H.W. Gardner, Advances in Lipid Methodology — Four (ed. W. W. Christie) The Oily Press, Dundee (1997), pp.1-43. A. Grechkin, Recent Developments in Biochemistry of the Plant Lipoxygenase Pathway, Progress in Lipid Research, 1998, 37, 317-352.

Lipoxygenase oxidation

liquid crystals
these are ordered forms of polar lipids in which the acyl groups are non-ordered (melted). The liquid crystalline state, also called mesomorphic state, may be induced by heat (thermotropic liquid crystals) or in the interaction with water (lyotropic liquid crystals). The latter form several phases (see hexagonal phase, lamellar phase, cubic phase).
K. Larsson, Lipids — Molecular Organisation, Physical Functions and Technical Applications, The Oily Press, Dundee, (1994).

liver oils
fatty extracts of animal liver. The most important are the fish liver oils which contain valuable *(n–3) fatty acids* and are a rich source of *vitamins A and D.*

long-chain bases
see *sphingoid bases.*

Lovibond tintometer
an instrument used to measure the colour of oils. Light transmission through a cell containing oil is compared with light through coloured slides. Results are generally quoted in yellow and red units.

Lovibond value
a system to express the colours of refined oils and fats. Red and yellow glasses are used, of which the red glasses are standardized. Attempts to modernize the method by spectrophotometry have not been fully adopted. Standard methods are described by AOCS [C13b 45(89].

low calorie fats
see *structured fats.*

low-density lipoproteins
see *lipoproteins.*

low-energy fats
see *structured fats.*

low-fat spread
see *margarine.*

LPC
see *lysophospholipids.*

LPE
see *lysophospholipids*.

LTA$_4$, LTB$_4$, LTC$_4$
see *leukotrienes*.

lumequic acid
a C$_{30}$ monoene acid (21c-30:1) present in seed oils of *Ximenia* spp.

lung surfactant
see pulmonary surfactant.

LUV
see *liposomes*.

LXA$_4$, LXB$_4$
see *lipoxins*.

lyotropic mesomorphism
see *liquid crystals*.

lysolecithin
see *lysophospholipids*.

lysophospholipids
phospholipids are esters of glycerol containing two acyl chains and a phosphatidic acid moiety (see *phosphatidyl esters*). Compounds containing only a single acyl chain (usually in position *sn*–1) are lysophospholipids and are further classified, e.g. as lysophosphatidylcholine (LPC) and lysophosphatidylethanolamine (LPE).

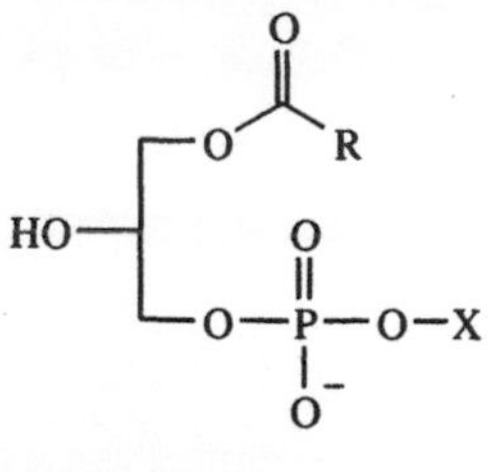

2-Lysophospholipids
LPC: X = choline
LPE: X = ethanolamine

M

macadamia oil
oil from macadamia nuts *(Macadamia phylla)*. An oleic-rich oil (ca 55%) which is also a convenient source of *9-hexadecenoic acid* (ca 22%). See also *speciality oils*.

MAG
see *monoacylglycerols*.

maize oil
a major vegetable oil. See *corn oil*.

malonaldehyde
see *malondialdehyde*.

malondialdehyde
a product of oxidation of polyenoic acids, able to cross link amino compounds. See *thiobarbituric acid*. Can also be measured directly by HPLC.
E.N. Frankel, Lipid Oxidation, The Oily Press, Dundee (1998).

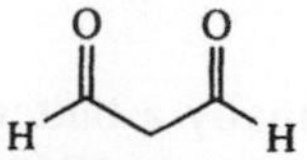

Malondialdehyde
Propandial
$C_3H_4O_2$
Mol. Wt.: 72.1

malonic acid
propanedioic acid (mp 133°C). An important *dibasic acid* in the acetate malonate biosynthetic pathway to fatty acids and other natural products. It is active as malonyl-CoA (mal-CoA).
J.L. Harwood, Lipid Synthesis and Manufacture (ed. F.D. Gunstone) Sheffield Academic Press, Sheffield (1999) p.422-466.

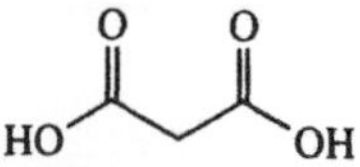

Malonic acid
Propandioic acid
$C_3H_4O_4$
Mol. Wt.: 104.1

malvalic acid

a C_{18} cyclopropene acid occurring with *sterculic acid* in the seed oil of the Malvales including *kapok seed oil* and *cottonseed oil*. Both these cyclopropene acids inhibit the biochemical desaturation of stearic acid to oleic acid. They are also called malvalinic acid and *halphenic acid*. They give a positive reaction in the *Halphen test*.

Malvalic acid
2-Octyl-1-cyclopropene-heptanoic acid
$C_{18}H_{32}O_2$
Mol. Wt.: 280.4

manaoic acid
see *cyclopentenyl acids*.

margaric acid
trivial name for *heptadecanoic acid*.

margarine
a water-in-oil emulsion, generally containing 80 % fat. Originally a butter substitute, invented as oleo-margarine by Méges-Mouriès in France in 1869, margarine is now a vegetable oil-based alternative to butter and animal fats and oils. Modern varieties, such as diet margarines and low-fat spreads, contain less fat and more water (down to 20% fat or below). The annual consumption of margarine (almost 10 million tonnes) now exceeds that of butter (almost 6 million tonnes).

E. Flack, Lipid Technologies and Applications (ed. F.D. Gunstone and F.B. Padley) Marcel Dekker, New York (1997), pp.305-327. M.M. Chrysam, Bailey's Industrial Oil and Fat Products, (ed. Y.H. Hui) John Wiley & Sons, New York (1996), Volume 3, pp.65-114.

marine oils
see *fish oils*.

mass detector
alternative name for the *evaporative light-scattering detector*.

mass spectrometry
mass spectrometry provides a method of structure identification. Molecules

bombarded with electrons give positively charged ions from the molecule or of its fragments. These are recognised in the spectrometer on the basis of their mass. This powerful detection technique can be coupled with gas chromatography (GC–MS) for the separation and identification of the individual components of a mixture. Olefinic acids are best identified as *picolinyl esters* or as *dimethyloxazoline* derivatives.

D.J. Harvey, *Advances in Lipid Methodology — One (ed. W. W. Christie) The Oily Press, Ayr (1992) pp.19-80. J-L Le Quere, Advances in Lipid Methodology — Two (ed. W. W. Christie) The Oily Press, Dundee (1993) pp.215-245. W.W. Christe, <www.lipid.co.uk/INFORES/index.html>.*

3-MCPD
see *3-chloropropanediol.*

MCT
see *medium-chain triacylglycerols.*

MDA
see malondialdehyde.

meadowfoam oil
the seed oil of *Limnanthes alba* with a very unusual fatty acid composition in that C_{20} and C_{22} acids make up 95% of the oil [5-20:1 (60–65%), 13-22:1 (~10%), and 5,13-22:2 (5–10%)]. Attempts are being made to grow this seed on a commercial basis. It has a number of promising non-food uses. The oil shows remarkable oxidative stability.

T.A. Isbell, *Development of meadowfoam as an industrial crop through novel fatty acid derivatives, Lipid Technology, 1997, 140-144. T.A. Isbell, Lipid Synthesis and Manufacture (ed. F.D. Gunstone) Sheffield Academic Press, Sheffield (1999) pp.401-421.*

Mead's acid
see *eicosatrienoic acid.*

medium-chain triacylglycerols
medium-chain triacylglycerols prepared industrially by esterification of glycerol with *octanoic* and *decanoic* acids (obtained from *coconut* and *palm kernel* oils). These esters are easily digested and are used for nutritional and pharmaceutical purposes. As liquids with high oxidative stability they also find use as food-grade lubricants.

F. Timmermann, *Production and properties of medium-chain triglycerides, Lipid Technology, 1994, 6, 61-64. J.W. Finley et al., Lipid Technologies and Applications (ed. F.D. Gunstone and F.B. Padley) Marcel Dekker, New York (1997), pp.501-510.*

Méges Mouriès
see *margarine.*

melissic acid
trivial name for triacontanoic acid (30:0, mp 93.6°C).

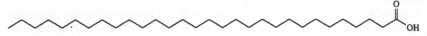

Melissic acid
Triacontanoic acid
$C_{30}H_{60}O_2$
Mol. Wt.: 452.8

melting point
(i) natural fats are complex mixtures of triglycerides with very many
molecular species. These mixtures do not exhibit a sharp melting point and
measurements must be obtained under closely defined conditions. In general
it is the temperature at which the fat becomes clear and liquid. See also
softening point, slip point. Standard methods are described by AOCS
[Cc 1 25(89) and Cc 2 38(89)].

(ii) defined lipid molecules such as fatty acids or glycerides may have several
melting points. See *polymorphism.*

membrane
biological membranes are based on bilayers mainly of phospholipids and
glycolipids. The properties of the bilayer are regulated by the polar head
groups and acyl chains of the molecules. Cholesterol molecules increase the
thickness of the membrane by causing straightening of the acyl groups. The
bilayers contain proteins, e.g. enzymes and receptors, which penetrate into or
through the bilayer and form the complete membrane structure.

membrane fluidity
variation in the composition of acyl groups of the *membrane lipids* causes
changes in the fluidity of the membrane. Differences in mobility and
temperature-induced transitions (from a viscous gel to a fluid state) are related
to the acyl chains. The mobility is also controlled by other components in the
membrane, e.g. proteins and cholesterol.

membrane lipids
general name for lipids that are present in biological membranes. The most
common are *phosphatidylcholines, phosphatidylethanolamines,*

phosphatidylinositols, phosphatidylserines, sphingomyelin, mono- and
digalactosyldiacylglycerols (in plants), *cholesterol* and *phytosterols.*

menhaden oil
an important commercial fish oil, produced in the USA.

mesomorphic state
synonymous to liquid crystalline state. See *liquid crystals.*

metathesis
a catalytic process involving exchange of alkyl groups attached to olefinic
carbon atoms. As an example, co-metathesis of ethene (ethylene) and methyl
oleate gives two C_{10} products, namely 1-decene and methyl 9-decenoate.

$CH_2=CH_2$ + $CH_3(CH_2)_7CH=CH(CH_2)_7COOMe$ → $CH_2=CH(CH_2)_7CH_3$ + $CH_2=CH(CH_2)_7COOMe$

*M. Sibeijn et al., Technologies and economical aspects of the metathesis of unsaturated esters, J. Am.
Oil Chem. Soc., 1994, 71, 553-561.*

methylene-interrupted polyene acids
most natural polyenoic acids contain two or more double bonds which have
cis configuration and are separated from each other by one methylene group.
The most important members belong to the n–3 or *n*–6 families. See, for
example, *octadecadienoic acid* and *octadecatrienoic acid.*
*J.-M.Vatele, Lipid Synthesis and Manufacture (ed. F.D. Gunstone) Sheffield Academic Press,
Sheffield (1999) pp.1-45.*

methanolysis
an example of *alcoholysis* in which the alcohol is methanol. For example,
reaction of mixed triacylglycerols with methanol and a catalyst (acidic, basic
or enzymic) gives methyl esters. This reaction is employed to produce methyl
esters on a small scale for gas chromatography and on a large scale for use as
biodiesel or for conversion to *alcohols.*
*W.W. Christie, Advances in Lipid Methodology — Two (ed. W. W. Christie) The Oily Press, Dundee
(1993) pp.69-111. F.D. Gunstone, Lipid Synthesis and Manufacture (ed. F.D. Gunstone) Sheffield
Academic Press, Sheffield (1999) pp.321-346.*

methyl ester
an ester derived from a (carboxylic) acid or derivative and methanol. Methyl
esters are made by *esterification* or *methanolysis.* They are the most
commonly used derivatives for *gas chromatography* and are produced on an
industrial scale for use as *biodiesel* or for conversion to *alcohols* and for other
purposes.

W.W. Christie, Advances in Lipid Methodology — Two (ed, W. W. Christie) The Oily Press, Dundee (1993) pp.69-111. F.D. Gunstone, Lipid Synthesis and Manufacture (ed. F.D. Gunstone) Sheffield Academic Press, Sheffield (1999) pp.321-346.

MG
see *monoacylglycerols.*

MGDG
see *monogalactosyldiacylglycerols.*

micelles
small colloidal aggregates formed in water by soluble *amphiphiles.* The micellar solutions exist above a certain concentration known as the *critical micellar concentration* (CMC). Typical compounds which form micelles are fatty acid salts (*soaps*), *lysophosphatidylcholines, monoacylglycerols,* and *bile salts.*

microbodies
see *peroxisomes.*

milk fat
see *butter fat.*

milk fat globule
see milk fat.

milk lipids
the lipid content of mammalian milk varies between approximately 2 and 10%. Cow milk is an oil-in-water emulsion, with 3 to 4% lipids of which ~98% consists of triacylglycerols (see *butter*). The rest is phospholipids and sphingolipids (approximately one-third each of *phosphatidylcholine, phosphatidylethanolamine* and *sphingomyelin*).

W.W. Christie, Advanced Dairy Chemistry: Lipids, Vol 2, 2nd edition: Composition and Structure of Milk Lipids, (ed. P.F. Fox) Chapman & Hall, London (1994), pp.1-36. P. Walstra, Advanced Dairy Chemistry: Lipids, Vol. 2, 2nd edition: Physical Chemistry of Milk Fat Globules (ed. P.F. Fox) Chapman & Hall, London (1994), pp.131-178. R.G. Jensen, The Lipids in Human Milk, Progress in Lipid Research, 1996, 35, 53-92. R.G. Jensen (ed.) Handbook of Milk Composition, Academic Press, San Diego, California (1995).

miscella
mixture of hexane, crude oil and water obtained during solvent extraction of vegetable oils.

mitochondrion
subcellular units or organelles in eukaryotic cells which are the major sites for beta-*oxidation* of fatty acids.

molecular species
individual molecules in a lipid class which differ due to the type or combination of acyl chains. The theoretical number of molecular species in a class, calculated from the total number of different acyl groups present (n) and the number of acyl groups in each molecule (x), is n^x. The triacylglycerols of natural oils may serve as an example: 10 acyl groups are combined on glycerol in triplets so there are 10^3 (1000) possible molecular species including all stereoisomers.

N.U. Olsson et al., Molecular species analysis of phospholipids, Journal of Chromatography B 692 (1997), 245-256.

monkey nut oil
see *groundnut oil.*

monoacylglycerols
a glycerol ester in which only one hydroxyl group has been esterified with a fatty acid. These exist in two forms depending on whether the primary (α-) or secondary hydroxyl (β-) is acylated. The unsymmetrical ester is chiral and there are two enantiomers depending on whether the acyl group is in the *sn*–1 or *sn*–3 position.
The pure isomers quickly change to a 9:1 mixture of the α- and β-isomers. This rearrangement is promoted by acid and by alkali. Monoacylglycerols and their derivatives with acetic acid (*Acetem^{TM}*), lactic acid (*Lactem^{TM}*), citric acid (*Citrem^{TM}*) or diacetyltartaric acid (*Datem^{TM}*) are used extensively as food emulsifiers. The monoacylglycerols themselves are designated E471 in Europe and S182 in USA. Mono- and di- acylglycerols are most commonly made from triacylglycerols by catalysed reaction with glycerol (*glycerolysis*). The product contains mono (45–55%), di (38–45%), and triacylglycerols (8–12%). Monoacylglycerols (93–97%) are obtained from this mixture by molecular distillation. Monoacylglycerols can also be made from glycerol and fatty acid with an enzymic catalyst.
Monoacylglycerols are formed in the intestines during digestion of *triacylglycerols* and are absorbed as such before being re-converted to triacylglycerols for transport as lipoproteins.
The 2-isomer exists in only one crystalline form but the 1-isomer exhibits in several crystalline forms, (*i.e.* they are polymorphic). Some 1- and 2-monoacylglycerols have the following melting points: monolaurin (63 and

51°C), monomyristin (70 and 60°C), monopalmitin (77 and 69°C), monostearin (77 and 74°C), monoolein (35 and 26°C), monoelaidin (58 and 54°C) and monolinolein (14 and 9°C).

N. Krog, Lipid Technologies and Applications (ed. F.D. Gunstone and F.B. Padley) Marcel Dekker, New York (1997), pp.521-534. G.L. Hasenheuttl, Lipid Synthesis and Manufacture (ed. F.D. Gunstone) Sheffield Academic Press, Sheffield (1999) pp.371-400.

1-isomer   2-isomer   3-isomer

Monoacylglycerol isomers

3-monochloropropanediol
see 3-*chloropropanediol.*

monoelaidin
see *monoacylglycerols.*

monoenoic acids
acids with one olefinic centre, generally having *cis (Z)* configuration. The most common members of this family are *hexadecenoic* (e.g. palmitoleic acid, 9*c*-16:1), *octadecenoic* (e.g. oleic acid, 9*c*-18:1 or 18:1(*n*–9), petroselinic acid, 6*c*-18:1), and *docosenoic acids* (e.g. erucic acid, 13*c*-22:1).

monogalactosyldiacylglycerols
lipid class present in plant membranes. See *galactosylglycerides, glycosylglycerides.*

monoglycerides
see *monoacylglycerols.*

monolaurin
see *monoacylglycerols.*

monolayer
normally refers to the monomolecular film of lipids on a water surface. The molecules are oriented with the polar head groups in the water and the acyl chains directed in the air. Most lipids form monolayers on water. See *Langmuir-Blodgett films*.

monolinolein
see *monoacylglycerols*.

monomyristin
see *monoacylglycerols*.

mono-olein
see *monoacylglycerols*.

monopalmitin
see *monoacylglycerols*.

monosialogangliosides
GM_1. See *gangliosides*.

monostearin
trivial name for glycerol monostearate which is used as a food surfactant and emulsifier, see also *monoacylglycerols*.

montanic acid
trivial name for octacosanoic acid (28:0, mp 90.5°C).

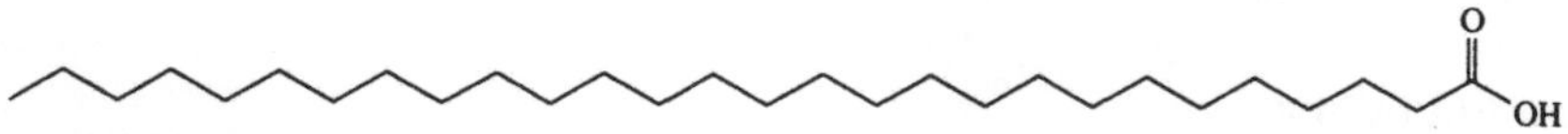

Montanic acid
Octacosanoic acid
(28:0)
$C_{28}H_{56}O_2$
Mol. Wt.: 424.7

moroctic acid, morotic acid
trivial names given to the *octadecatetraenoic acid* present in fish oils. It is probably 6,9,12,15-18:4 but was wrongly assigned the 4,8,12,15-18:4 structure. More commonly called stearidonic acid.

Mortierella alpina
a fungus producing lipid rich in arachidonic acid *(eicosatetraenoic acid)*.
See *Arasco™*.

moving-wire detector
see transport flame-ionization detector.

Mowrah butter
this fat from India and Sri Lanka comes from *Bassia latifolia* (or
Mowrah latifolia). With high levels of palmitic (24-28%), stearic
(14-19%) and oleic (43-49%) acids the fat is rich in SOS glycerol esters.
See *cocoa butter*.

Mowrah species
see *Mowrah butter*.

Mucor javanicus
the mould used for commercial production of an oil rich in γ-linolenic acid
(about 15%). The oil was produced and sold in the UK as `Oil of Javanicus' but
is no longer available. A similar product from *M. isabellina* was developed
in Japan.

Mucor species
see *lipases* and *Mucor javanicus*.

muscalure
see *pheromones*.

mutton tallow
see tallow.

myelin
the nerve-cell membrane with the highest lipid: protein ratio (3:1) of all
membranes. More than 40% of the lipids are *gangliosides*.

mycoceranic acid
2,4,6-trimethylhexacosanoic acid, also known as mycocerosic acid and
phthianoic acid. The 2R,4R,6R *isomer* occurs in the lipids of the
tubercle bacillus. (See structure on next page.)

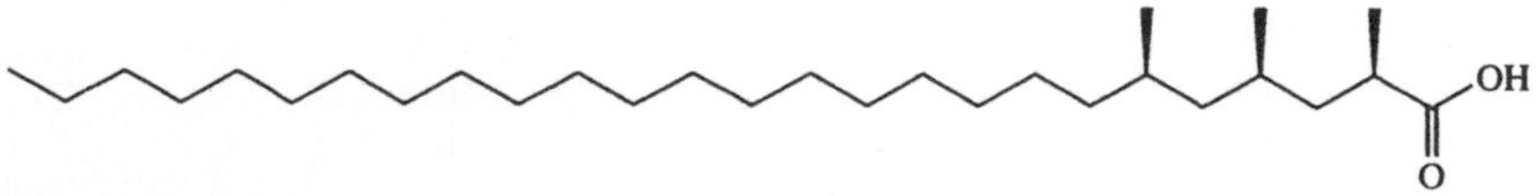

Mycoceranic acid
2*R*,4*R*,6*R*-Trimethylhexacosanoic acid
$C_{29}H_{58}O_2$
Mol. Wt.: 438.8

mycocerosic acid
see *mycoceranic acid.*

mycolic acids
α-branched-β-hydroxy acids of high molecular weight in mycobacteria. In the following structure R' is usually saturated. R may be saturated or unsaturated and may contain branched methyl groups, cyclopropane groups, and oxygenated functions. On pyrolysis they break down to RCHO and R'CH₂COOH.

$$RCHOHCH(R')COOH \rightarrow RCHO + R'CH_2COOH$$

C.E. Barry et al., Mycolic Acids: Structure, Biosynthesis and Physiological Functions, Progress in Lipid Research, 1998, 37, 2/3, 143-179.

myristelaidic acid
the *trans* isomer of myristoleic acid (9*t*-14:1, mp 18–18.5°C).

myristic acid
tetradecanoic acid, 14:0. This alkanoic acid is present in *coconut oil* and *palm kernel oil* (15–20%) and is a minor component of most animal fats and fish oils. It seems to be the principal saturated acid involved in raising plasma cholesterol levels.

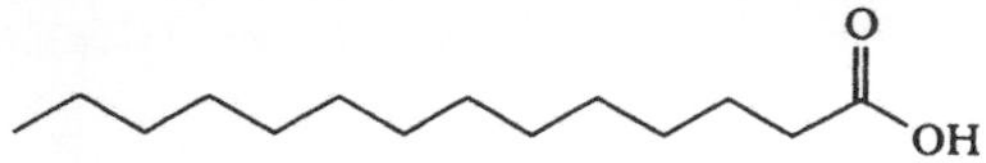

Myristic acid
tetradecanoic acid (14:0)
$C_{14}H_{28}O_2$ Mol. Wt.: 228.4

myristoleic acid
see tetradecenoic acid.

N

n-3 acids

also called ω3 (omega-3) acids. For the most part, these are a family of polyenoic acids with three or more *cis*-unsaturated centres separated from each other by one methylene group and having the first unsaturated centre three carbons from the end methyl. They are derived from α-linolenic acid, which is the first member, by chain-elongation and desaturation and include *eicosapentaenoic acid* (EPA) and *docosahexaenoic acid* (DHA). See also *essential fatty acids.*

n-3 acids

n-6 acids

also called ω6 (omega-6) acids. For the most part, these are a family of polyenoic acids with two or more *cis*-unsaturated centres separated from each other by one methylene group and having the first unsaturated centre on the sixth carbon from the methyl end. They are derived from *linoleic acid*, which is the first member, by chain-elongation and desaturation. Arachidonic acid (*eicosatetraenoic acid*) also belongs to this series. See also *essential fatty acids.*

n-6 acids

n-9 acids

(i) a series of *n*–9 (omega 9) monoene acids obtained from oleic acid by chain extension including *gondoic* (20:1), *erucic* (22:1), *nervonic* (24:1), ximenic (26:1), octacosenoic (28:1), and lumequic (30:1).

(ii) a series of methylene interrupted polyene acids with the first unsaturated centre nine carbons from the methyl end. They are derived from oleic acid by chain-elongation and desaturation. See, for example. *Mead's acid.* Also called ω9 (omega 9) acids.

n-9 acids

NDGA
see *nordihydroguaiaretic acid.*

near infrared spectroscopy
spectra through the range 800–2500 nm have been used to determine the oil-content of seeds etc and to provide some information on the fatty acid composition of the derived oil. See also *Fourier-transform infrared spectroscopy.*

nervonic acid
cis-15-tetracosenoic acid [15-24:1, mp 40.5–41°C (*cis*) 66–67°C (*trans*)]. It is a constituent of the sphingolipids of nerve tissue and is present in honesty seed oil (*Lunaria biennis*) (~25%).

J. Sargent and K. Coupland, Applications of specialized oils in the nutritional therapy of demyelinating disease, Lipid Technology, 1994, 6, 10-14.

neutralisation
a refining process for removing free acid from crude oil by reaction with alkali. Soap is removed by centrifugation and can be acidified to give fatty acid. Free acids are also removed by *physical refining* which is now preferred. See also *soap stock, acid oil* and *alkali refining.*

D.A. Allen, Lipid Technologies and Applications (ed. F.D. Gunstone and F.B. Padley) Marcel Dekker, New York (1997), pp.137-167.   W. De Greyt and M. Kellens, Edible Oil Processing (eds W. Hamm and R.J. Hamilton) Sheffield Academic Press, Sheffield (2000), pp.90-94.

neutral lipids
These are non-polar lipids in contrast to *polar lipids.* Examples include triacylglycerols, sterols, and sterol esters. See also *simple lipids.*

Niemann Pick disease
one of several *sphingolipidoses.* Symptoms are liver and spleen enlargement and mental retardation and these are accompanied by accumulation of *sphingomyelin* arising from a deficiency of the enzyme *sphingomyelinase.*

NIOP
National Institute of Oilseed Producers. A US organisation based in San Francisco, California, and devoted to the commercial use of seeds and their products.

niosomes
non-ionic *liposomes* composed of diacyl or monoacyl polyglycerol or polyoxyethylene based lipids in mixtures with cholesterol.

nisinic acid
the trivial name of 24:6 in fish oils. The structure is probably 6,9,12,15,18,21-24:6. It is now recogised as an important member of the *n*–3 pathway:

$$18:3 \rightarrow 18:4 \rightarrow 20:4 \rightarrow 20:5 \rightarrow 22:5 \rightarrow 24:5 \rightarrow 24:6 \rightarrow 22:6$$

NIR
see *Fourier-transform infrared spectroscopy.*

NMID
see *non-methylene-interrupted polyenes.*

NMR
see *nuclear magnetic resonance.*

nonadecanoic acid
the saturated C_{19} acid (mp 69.4°C).

nonanoic acid
the C_9 acid (pelargonic, mp 12.5°C, b.p. 256°C) obtained as one product of the *ozonolysis* of oleic acid.

non-dairy cream
a commercial cream-like material which usually contains a hardened *lauric oil* in place of milk fat.
E. Flack, Lipid Technologies and Applications (ed. F.D. Gunstone and F.B. Padley) Marcel Dekker, New York (1997), pp.305-328.

non-ionic surfactants
surface-active compounds with one or more long-chain alkyl groups and a

polar head group usually rich in oxygen. Many of these are produced by ethoxylation of long-chain alcohols or amides etc.

G. Bognolo, Lipid Technologies and Applications (ed. F.D. Gunstone and F.B. Padley) Marcel Dekker, New York (1997), pp.633-694.

$$CH_3(CH_2)_n \diagup O \diagdown \diagup O \diagdown_m H$$

Non-ionic surfactant

non-methylene-interrupted polyenes

acids with two (or more) unsaturated centres which are not (or not all) methylene interrupted. Double bonds are separated by more than one methylene group. For example, in *columbinic acid* (5t9c12c-18:3) double bonds at positions 5 and 9 are separated by two methylene groups. Other fatty acids of this type are found in seed oils of Gymnosperms, in meadowfoam oil, and in sponges (*demospongic acids*).

non-polar lipids

see *neutral lipids* and *simple lipids*.

non-saponifiable

see *unsaponifiable*.

nordihydroguaiaretic acid

a natural antioxidant from the desert plant *Larrea divaricata* (creosote bush) with the structure:

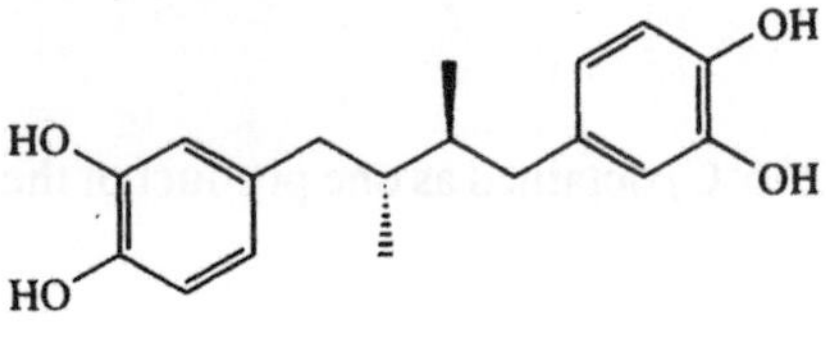

Nordihydroguaiaretic acid
4,4'-(2R,3S-Dimethyl-1,4-butanediyl)bis[1,2-benzenediol]
$C_{18}H_{22}O_4$
Mol. Wt.: 302.4

normal-phase chromatography

see *high-performance liquid chromatography*.

nuclear magnetic resonance

nuclear magnetic resonance exploits the nuclear spin of atoms such ^1H, ^{13}C, and ^{31}P. The spectra give the chemical shifts of atoms in different chemical environments and thereby provide valuable structural information. ^{31}P spectra are used for the study of *phospholipids* and ^1H and ^{13}C for the study of *fatty acids* and their *glycerol esters*.

F.D. Gunstone, Advances in Lipid Methodology — Two (ed. W. W. Christie) The Oily Press, Dundee (1993) pp.1-68. F.D. Gunstone, High resolution C-13 NMR. A technique for the study of lipid structure and composition, Progress in Lipid Research, 1994, 33, 1/2, 19-28. F.D. Gunstone, <www.lipid.co.uk/INFORES/index.html>.

NuSun

another oleic-rich (65%) sunflower oil with 9% saturated acids and 26% linoleic acid. It is expected that this will become the dominant form of sunflower oil in USA. See also *Sunola*.

M.K. Gupta, NuSun — healthy oil at a commodity price, Lipid Technology, 2000, 12, 29-33.

nylon-11

a polyamide, also called *Rilsan*, made from 11-amino-undecanoic acid and used as an engineering plastic. The amino acid is made from *10-undecenoic acid* produced by pyrolysis (450–500°C) of *castor* methyl esters.

O

obesity
normally refers to the human condition resulting from excessive storage of triacylglycerols in *adipose tissue*. A problem of growing concern in many developed countries.

T. Krawczyk, The spreading of obesity, INFORM, 2000, 11, 160-171.

obtusilic acid
see 4-*tetradecenoic acid.*

OCL
see *Oleagineux Corps gras Lipides.*

octadecadienoic acids
(i) the most important octadecadienoic acid (18:2) is linoleic acid (9*c*12*c* isomer) which is present in virtually all seed oils. It attains high levels in *sunflower* (48–74), *soybean* (50–57), *cottonseed* (47–58), *groundnut* (36–67), and *corn* or maize (40–62%) and is sometimes accompanied by *linolenic acid* (e.g. soybean oil 6–10%) It is the first member of the *n*–6 family of polyene acids and is an important *essential fatty acid.*

Linoleic acid
9Z,12Z-Octadecadienoic acid
$C_{18}H_{32}O_2$
Mol. Wt.: 280.4

(ii) *stereomutation* of linoleic acid gives a mixture of 9*t*12*t*, 9*c*12*t*, 9*t*12*c*, and 9*c*12*c* isomers. The 9*t*12*t* isomer (linelaidic acid) melts at 28–29°C.

(iii) dehydration of *castor oil* gives *dehydrated castor oil* (DCO). This contains several 18:2 acids including the 9*c*12*t* and 9*c*11*t* isomers. See also *conjugated linoleic acid.*

(iv) 5,9-octadecadienoic acid (*cis, cis* isomer), also known as taxoleic acid, is present in many conifer seed oils. Seeds of the yew (*Taxus baccata*) contain 10% of this acid which occurs almost entirely in the *sn*–3 position of triacylglycerols. See also *sciadonic, pinolenic* and *juniperonic acids.*

octadecanoic acid
the saturated C_{18} acid commonly known as stearic acid (mp 70.1°C). This is the second most common saturated acid after hexadecanoic (palmitic) and is produced commercially by hydrogenation of unsaturated C_{18} acids (oleic, linoleic, linolenic etc). It is a minor component of most vegetable oils but is present in larger amounts in vegetable butters such as *cocoa butter* (30–36%) and in *ruminant fats* (5–40%). Biosynthetically it is the precursor of oleic acid and thereby of almost all unsaturated acids.

Stearic acid
octadecanoic acid
(18:0)
$C_{18}H_{36}O_2$
Mol. Wt.: 284.5

octadecatetraenoic acid
(i) stearidonic acid (moroctic) is a member of the n–3 family ($6c9c12c15c$) and an intermediate in the metabolism of *α-linolenic acid* to acids such as *eicosapentaenoic acid* and *docosahexaenoic acid*. It is present in most fish oils and in some seed oils [e.g. blackcurrant (2–4%), *Echium plantagineum* (12–15%), *Onosmodium hispidissum* (~8%), and *Primula* spp (11–14%)]. It is now available commercially as glycerol esters with 6 and 14% of stearidonic acid. The acid inhibits release of PGE_2 and is a powerful moderator of inflammation including that induced by ultraviolet irradiation of the skin.

(ii) parinaric acid, a C_{18} acid with a conjugated tetraene system. The natural α-form ($9c11t13t15c$-18:4, mp 73°C), present in *Parinarium laurinum*, is readily transformed to the all-*trans* β-form (mp 95°C). See also *conjugated unsaturation*.

(iii) two non-methylene-interrupted polyenes are known ($3t9c12c15c$ and $5c9c12c15c$).

(iv) the *n*-4 acid ($5c8c11c14c$) is an arachidonoyl-CoA inhibitor.

octadecatrienoic acids
(i) The most common and important octadecatrienoic acid is α-linolenic acid which is the $9c12c15c$ isomer. This is a major component acid in *drying oils* such as *linseed* (50–60%) and of the glycolipids in photosynthetic tissue. It is

the first member of the *n–3 family* of polyene acids and is an *essential fatty acid*. When exposed to high temperature during *refining* the *cis* triene acid undergoes *stereomutation* to give mainly the 9*t*12*c*15*c* and 9*c*12*c*15*t* which are small but significant components of refined *soybean oil* and *rapeseed oil*.

α-Linolenic acid
9Z,12Z,15Z-Octadecatrienoic acid
$C_{18}H_{30}O_2$
Mol. Wt.: 278.4

(ii) γ-Linolenic acid, also called GLA, is a trivial name for 6*c*9*c*12*c*-octadecatrienoic acid. This is an isomer of the more common *α-linolenic acid*. It comes from a few plant sources and is available as a dietary supplement in the form of·*evening primrose oil, borage* (starflower) *oil*, and *blackcurrant seed oil*. It is a member of the *n–6* family of polyenes and an intermediate in the bio-conversion of *linoleic acid* to arachidonic acid (see *eicosatetraenoic acid*).

9,12-18:2 → 6,9,12-18:3 → 8,11,14-20:3 → 5,8,11,14-20:4

Claims have been made that it is beneficial in the treatment of several disease conditions. These claims are based on the fact that the first step in the above sequence is rate-limiting and that it is impaired in a variety of circumstances such as diabetes and advancing age.

F.D. Gunstone, Gamma linolenic acid — occurrence and physical and chemical properties, Progress in Lipid Research, 1992, 31, 145-161. D.F. Horrobin, Nutritional and medical importance of gamma-linolenic acid, Progress in Lipid Research, 1992, 31, 163-194. γ-Linolenic Acid — Metabolism and its Roles in Nutrition and Medicine, Y.-S. Huang and D.E. Mills (eds), AOCS Press, Champaign, Illinois (1995).

γ-Linolenic acid
6Z,9Z,12Z-Octadecatrienoic acid
$C_{18}H_{30}O_2$
Mol. Wt.: 278.4

(iii) Two 5,9,12 isomers occur naturally. The *cis*-5 isomer (pinolenic acid) is present in *tall oil* and in seed oils of many conifer species. The *trans*-5 isomer (columbinic acid) occurs in the seed oil of the columbine (aquilegia).

octadecenoic acids
the general name for any 18:1 acid including oleic, elaidic, vaccenic, petroselinic and petroselaidic.

(i) the *cis*-9 isomer (oleic, mp 12°C, 16°C) is the most common of all fatty acids and is present in almost every lipid. It serves biosynthetically as the precursor of the *n*–9 family of monoenes (such as *erucic* and *nervonic*) and the *n*–9 family of polyenes (such as *Mead's acid*). It is a major component in many oleic-rich oils including *olive* (55–83%), *almond* (65–70%), *macadamia* (50–59%), *NuSun* (65%), *low-erucic rape* (52–67%), and *groundnut* (>35%).

High-oleic *sunflower* (~80%) and *safflower* (~74%) have been developed. Newer oils inlcude high-oleic soybean (80%), canola (75%) and groundnut (80%). The oleochemical industry uses concentrates of oleic acid obtained from *tallow* or from *palm oil* by *hydrophilisation*. The *trans* isomer is called elaidic acid (mp 45°C).

Oleic acid
9Z-Octadecenoic acid
$C_{18}H_{34}O_2$
Mol. Wt.: 282.5

(ii) petroselinic (mp 33°C) and petroselaidic (mp 53°C) acids are the *cis* and *trans* isomers of 6-octadecenoic acid. The *cis* isomer occurs at high levels in most seed oils of the Umbelliferae family (e.g. carrot, parsley, coriander). Attempts are being made to develop coriander as a commercial seed oil and, by genetic engineering, to produce a *rapeseed oil* rich in this acid. *Ozonolysis* of this acid furnishes *adipic* and *lauric* acids.

Petroselinic acid
6Z-Octadecenoic acid
$C_{18}H_{34}O_2$ Mol. Wt.: 282.5

Petroselaidic acid
6E-Octadecenoic acid
$C_{18}H_{34}O_2$ Mol. Wt.: 282.5

(iii) vaccenic acid occurs in both *cis* and *trans* forms. The *cis* isomer (mp 15°C), also known as asclepic acid, is a minor component of most animal and vegetable fats. It is probably formed from 9-*hexadecenoic acid* by chain elongation. The *trans* isomer (mp 44°C) is the major acid in a group of *trans* octadecenoic acids present in *ruminant fats* and resulting from bio-hydrogenation of linoleic acid in the rumen. Along with other 18:1 isomers it is also a significant component in partially hydrogenated vegetable oils rich in linoleic acid.

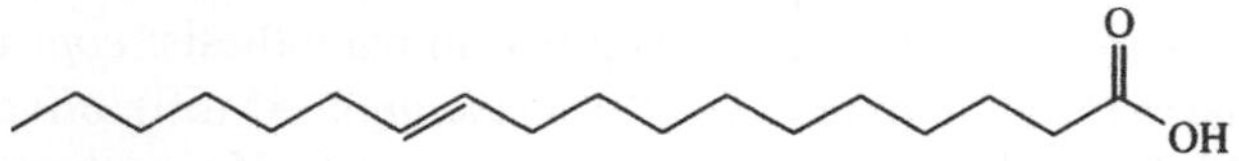

Vaccenic acid
11*E*-Octadecenoic acid
$C_{18}H_{34}O_2$
Mol. Wt.: 282.5

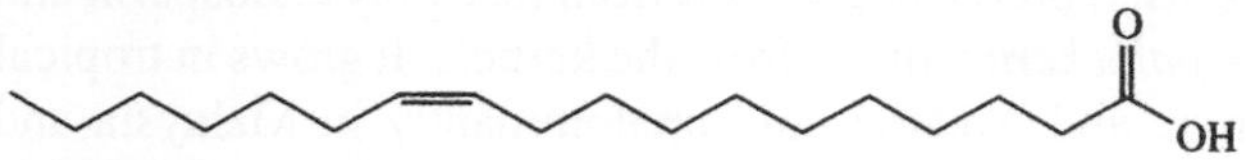

Aslepic acid
11Z-Octadecenoic acid
$C_{18}H_{34}O_2$
Mol. Wt.: 282.5

octanoic acid
the C_8 acid (caprylic, mp 16.5°C, bp 240°C). A medium-chain fatty acid which occurs in the *lauric oils* and in *Cuphea* seed oils. It is also a minor component of milk fats. It is a major acid in *medium-chain triglycerides* and in *caprenin*.

Oenothera biennis

see *evening primrose oil.*

Oenotherol ™
name given to the glycerol esters from *evening primrose oil* which contain two linoleic acid chains and one γ-linolenic acid chain.

Oenothera lamarkiana (Onagraceae)
see *evening primrose oil.*

Oil and Soap
earlier name for the *Journal of the American Oil Chemists' Society.*

oil bodies
sometimes called oleosomes, oleosins or lipid bodies. Cellular organelles in cells of oil seeds, containing mainly triacylglycerols.
D.J. Murphy, Storage lipid bodies in plants and other organisms, Progress in Lipid Research, 1990, 29, 299-324.

oil content
the quantity of oil in plant sources (seed or endosperm) varies with the source of material and with the extraction procedure (*cold pressing, solvent extraction*). The oil content of the major oils is given in parenthesis: *copra* (65–68%), *corn* or maize (5), *cottonseed* (18–20), *groundnut* (45–50), *olive* (25–30), *palm fruit* (45–50), *palm kernel* (45–50), *rapeseed* (40–45), *soybean* (18–20), and *sunflower* (35–45).

oil palm
this tree (*Elaeis guineensis*) produces *palm oil* from its fleshy endosperm and a quite different oil — *palm kernel oil* — from the kernels. It grows in tropical regions of Asia, Africa, and America and predominantly in Malaysia and Indonesia. At about 4 tonnes per hectare of the two oils combined on well-managed plantations, the oil palm outcrops all other oil crops. Fruit bunches of 4–20 kg each contain 200–2000 individual fruits which furnish palm oil (20–24%) and palm kernel oil (2–4%). Through seed breeding palm trees of lower height, higher oil yields, more unsaturated oil, and a higher proportion of kernel are becoming available.
B.S. Jalani and N. Rajanaidu, Improvements in oil palm: yield, composition, and minor components, Lipid Technology, 2000, 12, 5-8. Yusof Basiron, Palm oil, INFORM, 2000, 11, 30-33.

oil refining
industrial technology to obtain edible oils from crude oils through processing steps such as *degumming, neutralization, bleaching,* and *deodorization.*

oils
see *fats.*

Oils and Fats International
a business magazine for the oils and fats industry published by DMG Business Media Ltd (Redhill, UK).

oilseed extraction
method to produce crude oil from seeds. The seeds are crushed and/or flaked before entering the extractor where they are treated with industrial hexane.

Soybeans are the major oilseed processed by solvent extraction alone. Others, like sunflower and rapeseed, are pressed before the extraction.

M.A. Williams, Lipid Technologies and Applications (ed. F.D. Gunstone and F.B. Padley) Marcel Dekker, New York (1997), pp.113-135. J.-M. Fils, Edible Oil Processing (eds W. Hamm and R.J. Hamilton) Sheffield Academic Press, Sheffield (2000), pp.47-78.

Oil World

publications of a company (ISTA Mielke GmbH) in Hamburg, Germany, (established 1954) providing market data on 17 major oils and fats covering *soybean, cotton, groundnut, sunflower, rapeseed, sesame, corn, olive, palm, palm kernel, coconut, butter, lard, fish, linseed, castor and tallow*/grease. Weekly, quarterly, and annual publications are produced.

Oily Press

imprint of a book publishing company (P.J. Barnes & Associates, Bridgwater, UK) specialising in books on lipid analysis and other aspects of lipid science. Formerly The Oily Press Ltd based in Ayr and then Dundee. The publisher of this book.

Olea europaea (Oleaceae)

see *olive oil.*

Oleagineux

French journal on oils and oilseeds published since 1946 but now incorporated into *Oleagineux Corps gras Lipides.*

Oleagineux Corps gras Lipides

a French journal published since 1994 devoted to lipid science and technology and replacing older French journals.

oleic acid

trivial name for *cis-9-octadecenoic acid* (18:1). See *octadecenoic acids.*

oleic oils

oleic acid is usually obtained from *tall oil* ~(45%), *palm oil* ~(40%), or *tallow* ~(40%). Concentrates (70–75%) can be obtained from these relatively cheap sources by *hydrophilisation*. Richer sources of oleic acid include *rapeseed oil* (56%), *macadamia oil* (56% and a further 22 % of 16:1), *almond oil* (61%), *high-oleic safflower oil* (74%), *olive oil* (78%), high-oleic *sunflower oil* (*Sunola*, 82% and *NuSun*, 65 %) and *Euphorbia lathyris seed oil* (84%).

Newer oils inlcude high-oleic soybean (80%), canola (75%) and groundnut (80%).

olein
liquid fraction obtained by *fractionation* of, for instance, palm oil (see *stearin*).

oleochemicals
compounds produced on a commercial scale from natural oils and fats. They include *fatty acids, alcohols, methyl* or other *esters*, amides and *amines, dimer acids* and *dibasic acids* and are the basis of surfactants and other compounds.

oleosins
see *oil bodies*.

oleosomes
see *oil bodies*.

Olestra™
the trade name given to a sucrose polyester with 6–8 acyl groups per sucrose unit. It is made from sucrose and the fatty acids of soybean, corn, cottonseed, or sunflower oil. Olestra is not metabolised and has a virtually zero caloric value. It can be used as a frying oil, or as a replacement for fat in ice-cream, margarine, cheese and baked goods. It is non-toxic and non-carcinogenic. Permission has been given only for its limited use but this may be extended. Olestra™ absorbs fat-soluble *vitamins* and carotenoids and can cause loose stools and anal leakage.
J.W. Finley, A. McDonald et al., Lipid Technologies and Applications (ed. F.D. Gunstone and F.B. Padley) Marcel Dekker, New York (1997), pp.501-520. Sakidia and B.G. Swanson, Lipid Synthesis and Manufacture (ed. F.D. Gunstone) Sheffield Academic Press, Sheffield (1999) pp.347-370.

Olibra™
the trade name given to a lipid-based food ingredient which provides extended satiety compared to milk fat. It is a lipid extract dispersed in water which is made from *palm oil* and *oat* flakes and it can, for example, be consumed in a yoghurt. Olibra ™ contains only natural lipid components and has the same caloric value as other vegetable fats and oils.
Anon., Lipid Technology Newsletter 98, 4, 46; Lipid Technology, 98, 10, 3.

olive oil

a major vegetable oil obtained from the mesocarp of the fruits of the olive tree (*Olea europaea*). Annual production is about 2.5 million tonnes and commercial growth of the tree is confined almost entirely to Mediterranean countries (Italy, Greece, Spain, Turkey, and Tunisia). Virgin olive oil is produced from the first pressing and other grades of lower quality are subsequently produced. The oil is characterised by a high level of oleic acid with *codex* ranges of 8–20% for palmitic acid, 55–83% for oleic acid and 4–21% for linoleic acid. The major triacylglycerols are typically OOO (43), LOO (11), POO (22%). Olive oil contains *squalene* at a higher level (150–170 mg/100 g) than most vegetable oils (5–50 mg/100 g).

D. Firestone, E. Fedeli, et al., Bailey's Industrial Oil and Fat Products, (ed. Y.H. Hui) John Wiley & Sons, New York (1996), Volume 2, pp.241-269. D. Boskov, Olive oil: chemistry and technology, AOCS Press, Champaign, Illinois, 1996.

olive oil (grades)

oil obtained from the fruit of the olive tree only by mechanical or physical means under conditions, particularly thermal, that do not lead to alteration of the oil and which has not undergone treatment other than washing, decantation, centrifuging and filtration is called virgin olive oil. Extra virgin olive oil has perfect flavour and odour with maximum acidity of 1% (as oleic acid). Fine virgin oil also has perfect flavour and odour with maximum acidity of 2%. Semi-fine or ordinary virgin oil has good flavour and odour and maximum acidity of 3.3% with a 10% margin of tolerance. Virgin olive oil with an off-flavour or off-odour and acidity >3.3% is designated lampante. Refined olive oil, obtained from virgin olive oil by refining methods which do not affect fatty acid or glycerol ester composition, should have acidity <0.5%. Extraction of olive pomace gives a crude pomace oil. No oil can be added to this, nor should the product be esterified. It can be refined to an acidity <0.5% (refined olive-pomace oil) or mixed with virgin oil to give a mixture with acidity below 1.5%.

R. Aparicio, Characterization of virgin olive oils by SEXIA expert system, Progress in Lipid Research, 1994, 33, 1/2, 29-38. D. Firestone, E. Fedeli, et al., Bailey's Industrial Oil and Fat Products, (ed. Y.H. Hui) John Wiley & Sons, New York (1996), Volume 2, pp.241-269. D. Boskov, Olive oil: chemistry and technology, AOCS Press, Champaign, Illinois, 1996.

omega-3 acids

see *(n–3) acids.*

omega-6 acids

see *(n–6) acids.*

Omnium oxidative stability instrument

a machine for the measurement of resistance to oxidative rancidity. Results obtained at these elevated temperatures must be interpreted with care when using them to predict shelf life since the mechanisms of oxidation change with temperature.

E.N. Frankel, Lipid Oxidation, The Oily Press, Dundee (1998).

onagre (huile de)

French name for *evening primrose oil.*

oncobic acid

see *cyclopentenyl acids.*

ONIDOL

Organisation Nationale Interprofessionelle de Oleagineux. A French organisation founded in 1975 which groups all the partners in the French oilseed sector (mainly rapeseed, soybean and sunflower).

OPO

See *glycerol 1,3-dioleate 2-palmitate* and *Betapol™.*

organoleptic test

evaluation of food products by a combination of taste (by mouth) and smell (by nose) by taste panels. Four basic tastes are perceived by the mouth: sweet, sour, bitter and salt.

K. Warner, Bailey's Industrial Oil and Fat Products, (ed. Y.H. Hui) John Wiley & Sons, New York (1996), Volume 1, pp.105-144.

oryzanols

see *rice bran oil.*

Osbond acid

trivial name for *docosapentaenoic acid* (DPA).

oxalic acid

ethanedioic acid [mp 189°C (anhydrous) 101°C(dihydrate)]

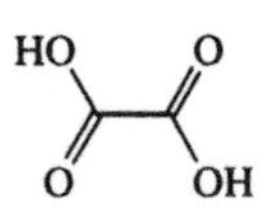

Oxalic acid
Ethanedioic acid
$C_2H_2O_4$
Mol. Wt.: 90.0

oxidation (α)

a biochemical mechanism of oxidation leading to α-hydroxy acids or acids with one less carbon atom (*nor*-acids). It occurs in plants, animals and microorganisms and is the source of some odd-chain aldehydes and acids. See *Refsum's disease*.

$$RCH_2COOH \longrightarrow RCH(OOH)COOH \longrightarrow RCHO \longrightarrow RCOOH$$

α-oxidation

oxidation (β)

a biological oxidation procedure occurring at the β (C3) position and leading to chain-shortening by two carbon atoms and eventually to complete catabolism by repetition of the process.

W.H Kanau et al., Beta-oxidation of fatty acids in mitochondria, peroxisomes, and bacteria: a century of continued progress, Progress in Lipid Research, 1995, 34, 267-342.

$$RCH_2CH_2COOH \longrightarrow RCH{=}CHCOOH \longrightarrow RCH(OH)CH_2COOH$$

$$\longrightarrow RCOCH_2COOH \longrightarrow RCOOH + CH_3COOH$$

β-oxidation

oxidation (ω)

a biochemical process resulting in oxidation of the ω-methyl group with production of a hydroxy acid, or dibasic acid. Oxidation can also occur at the penultimate carbon atom. For example enzymic oxidation of arachidonic acid (*eicosatetraenoic* acid) can occur at C20 or C19 to give a range of products including (the so-called) 20-carboxyarachidonic acid (5,8,11,14-eicosatetraenedioic acid), 20-hydroxyarachidonic acid, 19-hydroxyarachidonic acid, and 19-oxo-arachidonic acid.

$$CH_3(CH_2)_nCOOH \longrightarrow HOCH_2(CH_2)_nCOOH$$

ω-oxidation

oxidation by oxygen

unsaturated fatty acids are oxidised by oxygen under the influence of enzymes (*lipoxygenase* and *cyclo-oxygenase*) or without enzymes by *autoxidation* or *photo-oxygenation*. Other types of oxidation include *α-oxidation, β-oxidation, ω-oxidation, epoxidation, hydroxylation* and *oxidative cleavage* by ozone (*ozonolysis*) or other reagent. (References follow on the next page.)

E.N. Frankel, Lipid Oxidation, The Oily Press, Dundee (1998). D.B. Min, Food Lipids: Chemistry, Nutrition, and Biotechnology, (ed. C.C. Akoh and D.B. Min) Marcel Dekker, New York (1998) pp.283-296. M.C. Erikson, Food Lipids: Chemistry, Nutrition, and Biotechnology, (ed. C.C. Akoh and D.B. Min) Marcel Dekker, New York (1998) pp.297-332. J. Lawler and P.S. Dimick, Food Lipids: Chemistry, Nutrition, and Biotechnology, (ed. C.C. Akoh and D.B. Min) Marcel Dekker, New York (1998) pp.229-250. H. Zhuang, M.M. Barth et al. Food Lipids: Chemistry, Nutrition, and Biotechnology, (ed. C.C. Akoh and D.B. Min) Marcel Dekker, New York (1998) pp.333-375.

oxidative cleavage

oxidative fission of an acyl chain at an unsaturated centre by means of chemical reagents. See *olysis*, von *Rudloff oxidation*. The products of such reactions are usually a mixture of mono and dibasic acids and may provide evidence of the position of the unsaturated centre(s) in the acyl chain.

Oxidograph ™

an instrument for the accelerated measurement of induction period and the prediction of shelf life, based on oxygen-uptake. Suitable for fats and oils which are liquid at the test temperature. A modified instrument (Oxipres ™) can be used for multiphase and aqueous systems.

E.N. Frankel, Lipid Oxidation, The Oily Press, Dundee (1998).

Oxipres ™

see *Oxidograph ™*

oxo-octadecanoic acids

the whole series of 2-oxo to 17-oxo-octadecanoic acids have been synthesised. The 4- isomer and the 6-isomer (lactarinic acid) occur naturally; the 9-, 10-, 13, 15, 16, and 17-oxo acids are trace constituents of milk fats. The 12-isomer is easily made from *ricinoleic acid.*

ozonolysis

a method for oxidative cleavage of unsaturated acids or esters with ozone (to give an ozonide) followed by oxidative or reductive cleavage. Oleic acid furnishes a mixture of six ozonides. Their breakdown products may be acids, esters, aldehydes, alcohols or amines depending on the reagents employed. The reaction is used for structure-determination and for large scale preparative purposes. See also *oxidation, von Rudloff oxidation, dibasic acids.*

L. Rebrovic and F.D. Gunstone, Oxidative cleavage of unsaturated fatty acids, Lipid Technology, 1996, 8, 135-137.

P

PA
see *phosphatidic acid.*

PAF
see *platelet-activating factor.*

palm fatty acid distillate
generally designated PFAD, this fraction is a by-product of the physical refining of *palm oil*. It is rich in free acids and is used as a constituent of animal feed in the form if its calcium salt. In ruminants such salts are insoluble in the rumen and hence do not interfere with rumen metabolism but dissociate under the acidic conditions of the abomasum to give free acids.

palmiste (huile de)
French name for *palm kernel* oil.

palmitelaidic acid
trivial name for the *trans* isomer of 9-hexadecenoic acid (mp 32–33°C).

palmitic acid
trivial name for *hexadecanoic acid* (16:0).

palmitoleic acid
9-cis-*hexadecenoic acid*, 9c-16:1, also known as zoomaric acid. Present in fish oils (around 10%) and in small amounts in most animal and vegetable fats. See also 3-*trans-hexadecenoic* acid and *monoenoic acids.*

palm kernel oil
produced from the kernels of the *oil palm* usually by solvent extraction. It is quite different from *palm oil*. It is a *lauric oil*, similar in composition to *coconut oil* (lauric acid ~50% and myristic acid ~16%). It has a number of food and non-food uses, the latter mainly in the form of the corresponding alcohols. Annual production is approaching 2.3 million tonnes. See also *palm oil.*

palm mid fraction
this material (PMF) is a product of the *fractionation* of palm oil which can be

used as a *cocoa butter equivalent*. One route is the refractionation of *palm olein* into a *superolein* and a soft PMF (IV 42–48). Refractionation of this product gives an olein which can be recycled and a hard PMF (IV 32–36). Compared with cocoa butter the product is over-rich in POP and can be improved by enzymic *acidolysis* with stearic acid and a 1,3-specific enzyme.

R.E. Timms, *Lipid Technologies and Applications (ed. F.D. Gunstone and F.B. Padley) Marcel Dekker, New York (1997), p.199.*

palm oil
the oil pressed from the fleshy endosperm of the fruits of the *oil palm*. The supply of this oil has risen considerably since about 1980. It is expected to exceed 20 million tonnes in 2000 and should exceed the supply of *soybean oil* early in the next decade. The oil contains almost equal proportions of saturated (palmitic ~48% and stearic ~4%) and unsaturated acids (oleic ~37% and linoleic ~10%) and the major triacylglycerols are POP (30–40%), and POO (20–30%). The oil can be *fractionated* to give *palm stearin, palm olein*, and *palm mid fraction*. It is used mainly for food purposes but has some non-food uses. Valuable by-products obtained from palm oil are *carotene, tocopherols* and *tocotrienols (vitamin E)* and *palm-fatty acid distillate (PFAD)*. See also *red palm oil*.

L. deMan and J.M. deMan, *Functionality of palm oil in margarines and shortenings, Lipid Technology, 1994, 6, 5. Y. Basiron, Bailey's Industrial Oil and Fat Products, (ed. Y.H. Hui) John Wiley & Sons, New York (1996), Volume 2, pp.271-375. Yusof Basiron, Palm oil beyond 2000, INFORM, 2000, 11, 30-33.*

palm olein
palm oil (mp 21–27°C) is separated into *palm stearin* (30–35% of the original oil, mp 48–50°C) and palm olein (65–70%, mp 18–20°C). The latter finds a ready market as a high quality, highly stable, frying oil. With improved filtration procedures the yield of olein has been raised to 71–78%. It has a *cloud point* of 7–10°C and *iodine value* 57–59 and can be fractionated further to give *palm mid fraction* and a *superolein* (cloud point 3–4°C, IV 64–66) and a top olein (cloud point <0°C, IV 70–72). A typical palm olein has the following composition (compared to palm oil): palmitic 39 (44), oleic 44 (40) and linoleic 11% (10). The total level of saturated acids is reduced from 50% in palm oil to 45% in palm olein.

R.E. Timms, *Lipid Technologies and Applications (ed. F.D. Gunstone and F.B. Padley) Marcel Dekker, New York (1997), pp.199-222.*

palm stearin
palm oil (mp 21–27°C) is separated into palm stearin (30–35% of the original

oil, mp 48–50°C) and *palm olein* (65–70%, mp 18-20°C). The olein is the more valuable product but the stearin can be used as a hard fat in margarine stock or as an alternative to tallow in the oleochemical industry where it serves as a source of palmitic and oleic glycerides. Recrystallisation of the stearin (IV 32–36) gives soft stearin (IV 40–42) and superstearin (IV 17–21). These three products contain 68, 60, and 83% of saturated acids respectively.

R.E. Timms, Lipid Technologies and Applications (ed. F.D. Gunstone and F.B. Padley) Marcel Dekker, New York (1997), pp.199-222

pancreatic lipase
this lipase from the pancreas (pig is the main commercial source) selectively deacylates *triacylglycerols*, removing acyl groups from the *sn*–1 and *sn*–3 positions only. This change occurs in the body. It can also be used in the laboratory to obtain analytical information on the regiospecific distribution of fatty acids in triacylglycerols. See also *lipase* and *lipolysis*.

G.P. McNeill, Lipid Synthesis and Manufacture (ed. F.D. Gunstone) Sheffield Academic Press, Sheffield (1999) pp.288-320.

pancreatic lipases
see *lipases.*

parinaric acid
trivial name for isomers of the conjugated tetraene acid 9,11,13,15-octadecatetraenoic acid. See *conjugated unsaturated acids.*

partial glycerides
glycerol forms mono-, di- and triacyl esters. Natural fats are mainly *triacylglycerols* (triglycerides). Other glycerol esters are described collectively as partial glycerides (*mono*- and *diacylglycerols*).

partial hydrogenation
see *hydrogenation.*

partition chromatography
chromatography with a non-polar stationary phase, e.g. paraffin impregnated silica or silica with chemically-bonded C_{18} chains, and a polar mobile phase such as methanol-water mixtures. This type of chromatography separates lipids by the partitioning of the acyl chains and is used for the separation of *molecular species.*

partition number
normally refers to the relative retention of *triacylglycerols* in reversed-phase
partition chromatography. Unsaturated *molecular species* which coelute
with saturated compounds have the same partition number. Triolein and
tripalmitin, for example, traditionally both had partition number 48 but can
now be separated. See also *critical pairs*.

passionflower oil
a linoleic-rich oil (ca 75%) from the seed of the passionflower (*Passiflora
incarnata*). See also *speciality oils*.

PC
see *phosphatidylcholines*.

PE
see *phosphatidylethanolamines*.

peanut oil
see *groundnut oil*.

pecan oil
an oleic-rich oil (ca 65%) from pecan seeds (*Carya illinoensis*). See also
speciality oils.

pelargonic acid
see *nonanoic acid*.

Penicillium species
see *lipases*.

permanganate–periodate oxidation
see *von Rudloff oxidation*.

peroxide value
a measure of *hydroperoxides* in oxidised oil. These are measured
quantitatively on the basis of their ability to liberate iodine from acidic
solutions of potassium iodide. This can be measured by titrating with sodium
thiosulphate solution or electrochemically. The value is quoted as mmol of
available oxygen per 2 kg of fat. Freshly refined material should have a

peroxide value below 1. Fat is considered *rancid* by the time the peroxide value has reached 10. See also *anisidine value, totox value,* and *Fourier-transform near infrared spectroscopy.* Standard methods are described by AOCS [Cd 8 53(89)] and IUPAC (2.501).

E.N. Frankel, Lipid Oxidation, The Oily Press, Dundee (1998).

peroxisomes
cellular organelles in animal tissues which are sites for β-*oxidation.*

peroxy acids
acyl derivatives of hydrogen peroxide which act as epoxidising agents. They are made from the appropriate acid (e.g. lauric, myristic, palmitic and stearic acids) by interaction with hydrogen peroxide in the presence of an acidic catalyst.

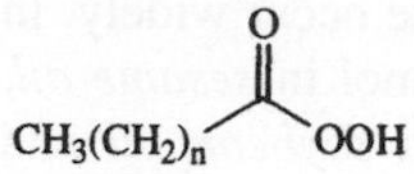

$$CH_3(CH_2)_n \overset{\displaystyle O}{\underset{}{C}} OOH$$

Peroxy acid

petroselaidic acid
trivial name for *trans-6-octadecenoic acid* (18:1).

petroselinic acid
trivial name for *cis-6-octadecenoic acid* (18:1).

PF
see *protection factor.*

PG
see *phosphatidylglycerols.*

PG
see *propyl gallate.*

PGE$_2$
see *prostaglandin.*

PGI$_2$
see *prostacyclin.*

PGH$_2$
see *prostaglandin H synthase*.

phellonic acid
trivial name for 22-hydroxydocosanoic acid (mp 100°C), present in cork.

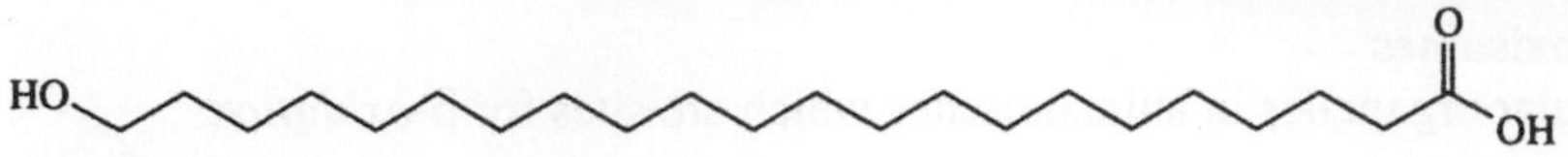

Phellonic acid
22-Hydroxydocosanoic acid
C$_{22}$H$_{44}$O$_3$
Mol. Wt.: 356.6

phenolic antioxidants
natural phenolic compounds that act as *antioxidants*. These occur widely. In some cases they are extracted with the oils such as sesamol in *sesame oil,* caffeic acid in soybean oil, *ferulic acid* in *corn oil* and *soybean oil* and oryzanols in *rice bran oil.* Many herbs and spices used to preserve foods contain phenolic compounds with antioxidant activity, e.g. thyme, sage, myrtle, tea, oats.

pheromones
insect attractants. A large number of chemicals act in this capacity and many are lipophilic in nature. They include aldehydes, alcohols, and their acetates and are related biosynthetically to fatty acids. One example is the hydrocarbon muscalure (*cis*-9-tricosene) which is a sex attractant for the housefly.

phloionolic acid
the 9*S*,10*S* form of 9,10,18-trihydroxystearic acid (mp104–105°C) is a constituent of cork.

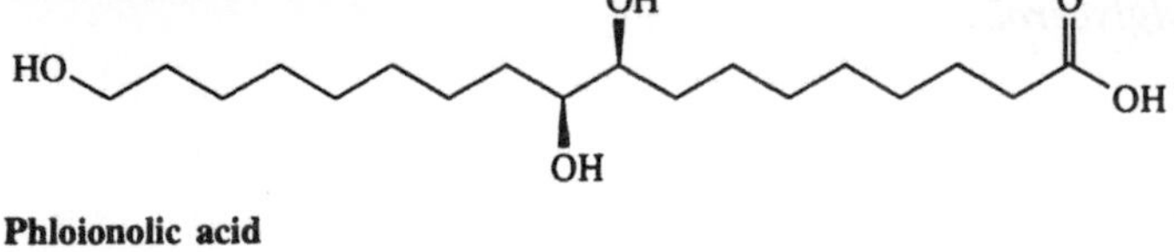

Phloionolic acid
9*S*,10*S*,18-Trihydroxyoctadecanoic acid
C$_{18}$H$_{36}$O$_5$
Mol. Wt.: 332.5

phosphatides
see *phospholipids*.

phosphatidic acids

PA; a negatively-charged (or acidic) class of phospholipids occurring only in low levels but important as intermediates in the biosynthesis and metabolism of other phospholipids.

R. Bittman, Lipid Synthesis and Manufacture (ed. F.D. Gunstone) Sheffield Academic Press, Sheffield (1999) pp.185-207.

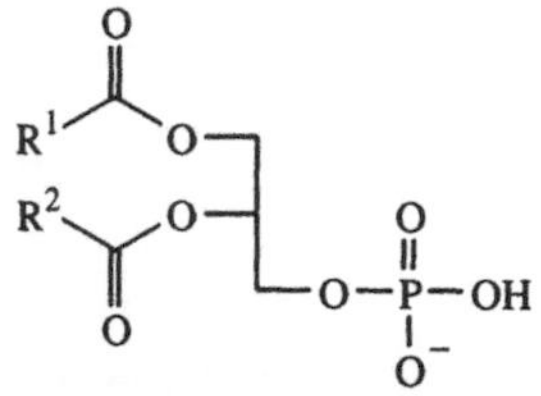

Phosphatidic acids

phosphatidylcholines

PC, a major class of phospholipids which are important components of the *membrane lipids* in animals and plants. They are zwitterionic and contain, in bound form, glycerol, phosphoric acid, fatty acids (2 moles) and choline. A major component of crude lecithin from soybean and other vegetable oils.

R. Bittman, Lipid Synthesis and Manufacture (ed. F.D. Gunstone) Sheffield Academic Press, Sheffield (1999) pp.185-207.

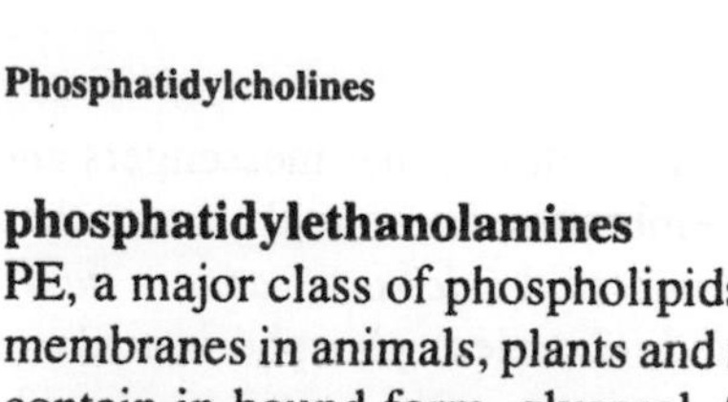

Phosphatidylcholines

phosphatidylethanolamines

PE, a major class of phospholipids which are important components of lipid membranes in animals, plants and microorganisms. They are zwitterionic and contain in bound form, glycerol, fatty acids (2 moles), phosphoric acid and ethanolamine. A major component of crude lecithin from soybean and other vegetable oils.

R. Bittman, *Lipid Synthesis and Manufacture* (ed. F.D. Gunstone) Sheffield Academic Press, Sheffield (1999) pp.185-207.

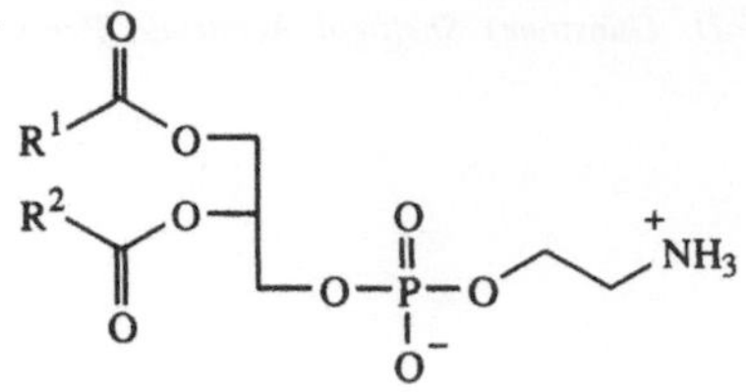

Phosphatidylethanolamines

phosphatidylglycerols

PG, a major class of negatively charged (or acidic) phospholipids, particularly important in the lung, in photosynthetic tissue, and in many bacteria. Complete hydrolysis gives glycerol (2 moles), fatty acids (2 moles) and phosphoric acid.

R. Bittman, *Lipid Synthesis and Manufacture* (ed. F.D. Gunstone) Sheffield Academic Press, Sheffield (1999) pp.185-207.

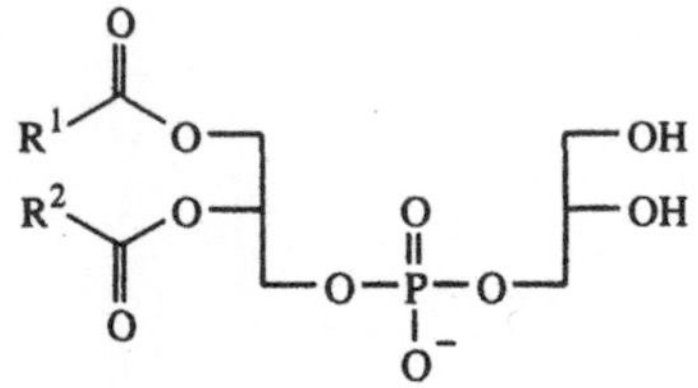

Phosphatidylglycerols

phosphatidylinositols

PI, a major class of phospholipids, important as intracellular messengers and protein anchor substances (*Lipid A*) in membranes. In animal tissues they often contain high proportions of arachidonic acid (*eicosatetraenoic acid*). Complete hydrolysis gives glycerol, fatty acids (2 moles), phosphoric acid and *myo*-inositol. See also *phosphoinositides*.

R. Bittman, *Lipid Synthesis and Manufacture* (ed. F.D. Gunstone) Sheffield Academic Press, Sheffield (1999) pp.185-207.

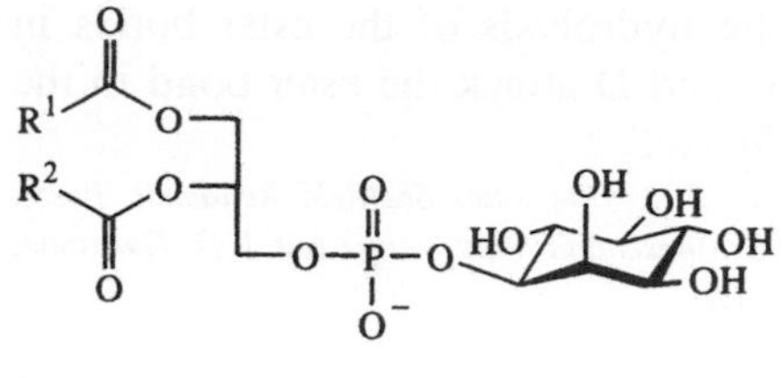

Phophatidylinositols

phosphatidylserines

PS, a class of phospholipids, widespread but minor. Important constituent of many tissue membranes, e.g. the brain. Complete hydrolysis gives glycerol, fatty acids (2 moles), phosphoric acid and L-serine.

R. Bittman, Lipid Synthesis and Manufacture (ed. F.D. Gunstone) Sheffield Academic Press, Sheffield (1999) pp.185-207.

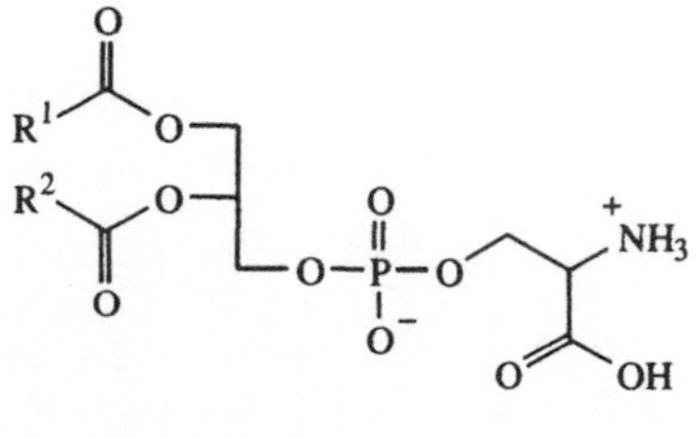

Phophatidylserines

phosphoglycerides

see *phospholipids*.

phosphoinositides

classes of phospholipids based on *phosphatidylinositol*. Normally 1 to 3 phosphate groups are attached to the inositol head group and participate in biochemical reactions. For example, phosphatidylinositol bisphosphate is hydrolysed by a special phospholipase C to release inositol 1,4,5-triphosphate and diacylglycerols, both of which having messenger functions in regulating cellular metabolism.

R. Bittman, Lipid Synthesis and Manufacture (ed. F.D. Gunstone) Sheffield Academic Press, Sheffield (1999) pp:185-207.

phospholipases
a group of enzymes which promote the hydrolysis of the ester bonds in phospholipids. Phospholipase A_1, A_2, C and D attack the ester bond in the positions shown. See also *sphingomyelinase*.

R. Bittman, Lipid Synthesis and Manufacture (ed. F.D. Gunstone) Sheffield Academic Press, Sheffield (1999) pp.185-207. G.P. McNeill, Lipid Synthesis and Manufacture (ed. F.D. Gunstone) Sheffield Academic Press, Sheffield (1999) pp.288-320.

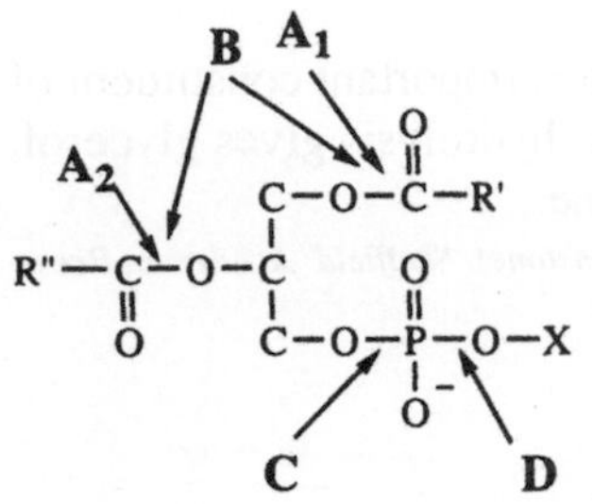

Phospholipases

phospholipase A_1
promotes the hydrolysis of phospholipids such as *phosphatidylcholines* to fatty acids (from the *sn*–1 position) and lysophosphatidylcholine.

R. Bittman, Lipid Synthesis and Manufacture (ed. F.D. Gunstone) Sheffield Academic Press, Sheffield (1999) pp.185-207. G.P. McNeill, Lipid Synthesis and Manufacture (ed. F.D. Gunstone) Sheffield Academic Press, Sheffield (1999) pp.288-320.

phospholipase A_2
promotes the hydrolysis of phospholipids such as *phosphatidylcholines* to fatty acids (from the *sn*–2 position) and lysophosphatidylcholine. Snake venom is the common source for practical preparations of the enzyme for use in the structural analysis of lipids.

R. Bittman, Lipid Synthesis and Manufacture (ed. F.D. Gunstone) Sheffield Academic Press, Sheffield (1999) pp.185-207. G.P. McNeill, Lipid Synthesis and Manufacture (ed. F.D. Gunstone) Sheffield Academic Press, Sheffield (1999) pp.288-320.

phospholipase B
promotes the hydrolysis of both acyl groups (*sn*–1 and *sn*–2) of phospholipids. It liberates fatty acids (2 moles) and produces glycerophosphorycholine. Most preparations, originally designated phospholipase B, were found to contain a mixture of the A_1 and A_2 phospholipases.
R. Bittman, Lipid Synthesis and Manufacture (ed. F.D. Gunstone) Sheffield Academic Press, Sheffield (1999) pp.185-207. G.P. McNeill, Lipid Synthesis and Manufacture (ed. F.D. Gunstone) Sheffield Academic Press, Sheffield (1999) pp.288-320.

phospholipase C
promotes the hydrolysis of phospholipids such as *phosphatidylcholines* to give 1,2-diacylglycerols and phosphorylcholine. Bacteria (e.g. *Bacillus cereus*) are a common source for bulk preparations for use in the structural analysis of lipids.
R. Bittman, Lipid Synthesis and Manufacture (ed. F.D. Gunstone) Sheffield Academic Press, Sheffield (1999) pp.185-207. G.P. McNeill, Lipid Synthesis and Manufacture (ed. F.D. Gunstone) Sheffield Academic Press, Sheffield (1999) pp.288-320.

phospholipase D
promotes the hydrolysis of phospholipids such as *phosphatidylcholines* to give *phosphatic acids* and choline. Plants, such as cabbage or carrots, are common sources.
R. Bittman, Lipid Synthesis and Manufacture (ed. F.D. Gunstone) Sheffield Academic Press, Sheffield (1999) pp.185-207. G.P. McNeill, Lipid Synthesis and Manufacture (ed. F.D. Gunstone) Sheffield Academic Press, Sheffield (1999) pp.288-320.

phospholipids
a general description of lipids containing phosphoric acid (or other phosphorus-containing acids) in appropriate ester form such as glycerophospholipids (e.g. *phosphatidic acid, phosphatidylcholine, phosphatidylethanolamine*) or sphingophospholipids (e.g. *sphingomyelin*). See also *phosphonolipids*.
M. Schneider, Lipid Technologies and Applications (ed. F.D. Gunstone and F.B. Padley) Marcel Dekker, New York (1997), pp.51-78. M.C. Erickson, Food Lipids: Chemistry, Nutrition, and Biotechnology, (ed. C.C. Akoh and D.B. Min) Marcel Dekker, New York (1998) pp.37-53.

phosphonolipids
phospholipids based on phosphonic acid, *i.e.* having a C–P bond. Natural examples include phosphono analogues of phosphatidylethanolamine and ceramide phosphorylethanolamine.

R. Bittman, *Lipid Synthesis and Manufacture (ed. F.D. Gunstone) Sheffield Academic Press, Sheffield (1999) pp.185-207.*

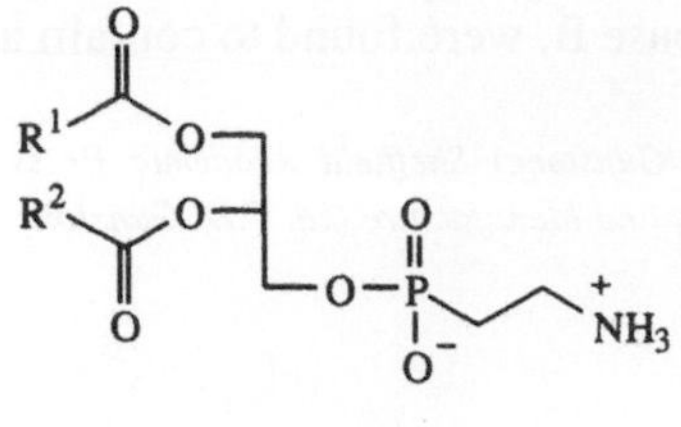

Glycerophosphonolipid
1,2-diacyl-sn-glycero-3-(2'-aminoethyl)phosphonate

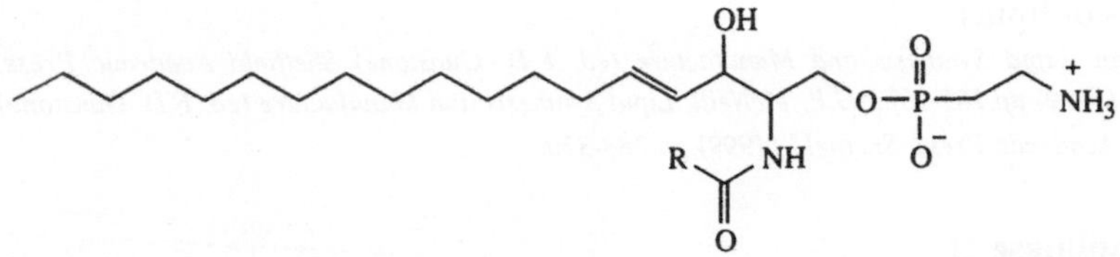

Sphingophosphonolipid
Ceramide-1-(2'-aminoethyl)phosphonate

photo-oxidation

this involves an ene reaction between an olefinic centre and singlet oxygen. This reactive form of oxygen is produced by light and a sensitiser such as chlorophyll, erythrosine, rose bengal, methylene blue, etc. The product is a *hydroperoxide* and reaction occurs with stereomutation and double bond migration. Unlike *autoxidation* there is no *induction period* and the reaction is unaffected by *antioxidants*. It can be inhibited by singlet oxygen quenchers such as *carotene*. The reaction is quicker than autoxidation.
E.N. Frankel, Lipid Oxidation, The Oily Press, Dundee (1998).

phrenosic acid

see *cerebronic acid.*

phrenosinic acid

see *cerebronic acid.*

phthianoic

see *mycoceranic acid.*

phthioic acid
an early name for the dextrorotatory acid(s) present in the tubercle bacillus. It was later shown to be a mixture of polybranched acids.

physeteric acid
the trivial name for 5-*cis*-tetradecenoic acid (14:1), a member of the *n*–9 family present in sperm whale oil and in lipids of some microorganisms.

Physeteric acid
5Z-Tetradecenoic acid
$C_{14}H_{26}O_2$
Mol. Wt.: 226.4

physetoleic acid
see *hexadecenoic acid.*

physical refining
a refining procedure for removing mono- and diacylglycerols, fatty acids, oxidation products and pigment-decomposition products from an oil. The oil is heated at temperatures up to 260°C under reduced pressure and with steam injection. This method is only suitable for oils with a low content of phospholipids. At the high temperature involved in this process a proportion of the polyene acids, particularly linolenic, is converted to *trans* isomers.

D.A. Allen, Lipid Technologies and Applications (ed. F.D. Gunstone and F.B. Padley) Marcel Dekker, New York (1997), pp.137-167. W. De Greyt and M. Kellens, Edible Oil Processing (eds W. Hamm and R.J. Hamilton), Sheffield Academic Press, Sheffield (2000), pp.105-121.

phytanic acid (phytanoic acid)
trivial name for 3,7,11,15-tetramethylhexadecanoic acid. This an isopropenoid acid resulting from the oxidation of *phytol* which is a component of chlorophyll. The acid is present in many land and marine animal fats. See also *Refsum's disease.*

Phytanic acid
3*R*,7*R*,11*R*,15-Tetramethyldecanoic acid
$C_{20}H_{40}O_2$
Mol. Wt.: 312.5

phytenoic acid
the monounsaturated C_{20} branched chain acid resulting from the corresponding alcohol phytol.

Phytenoic acid
3,7R,11R,15-Tetramethyl-2E-hexadecenoic acid
$C_{20}H_{38}O_2$
Mol. Wt.: 310.5

phytoceramide-1
a derivative of ω-hydroxyheptacosanoic acid (C_{27}) in which the hydroxyl groups is esterified with stearic acid and the carboxyl group is linked as amide to *phytosphingosine.*

phytol
a C_{20} terpene alcohol present in bound form in chlorophyll. It is the precursor of several polybranched-chain acids present in some animal fats.

Phytol
3,7R,11R,15-Tetramethyl-2E-hexadecen-1-ol
$C_{20}H_{40}O$
Mol. Wt.: 296.5

phytomonic acid
see *lactobacillic acid.*

phytosphingosine
4-hydroxysphinganine. A long-chain base present in *sphingolipids* and mainly of plant origin. It can be produced from palmitic acid and L-serine by fermentation.

phytosterols
plant derived sterols The most important phytosterols in soybean are *campesterol, stigmasterol* and *sitosterol.* About 75% of the world supply of corticosteroids and sex hormones is now produced from the mixed sterols recovered from soybean oil in the *deodorizer distillate.* They are added to

spreads to reduce cholesterol absorption. See *Benecol*[TM].

J.P. Clark Tocopherols and sterols from soybeans,, Lipid Technology, 1996, 8, 111-114.

PI
see *phosphatidylinositol.*

picolinyl esters
made by reacting acid chlorides with 3-pyridylcarbinol. These derivatives are used in GC–MS studies to determine double bond position in unsaturated acids and location of other functional groups.

D.J. Harvey, Advances in Lipid Methodology — One (ed. W.W. Christie), The Oily Press, Ayr, 1992, pp.19-80.   W.W. Christie, Beginner's guide to mass spectrometry of fatty acids. General purpose derivatives, Lipid Technology, 1996, 8, 64-66.   W.W. Christie, <www.lipid.co.uk/INFORES/ index.html>. F.D. Gunstone, Lipid Synthesis and Manufacture (ed. F.D. Gunstone), Sheffield Academic Press, Sheffield (1999) pp.321-346.

pilchard oil
see sardine oil.

pimelic acid
trivial name for the C_7 dibasic acid (heptanedioic, mp 105°C).

Pimelic acid
Heptanedioic acid
$C_7H_{12}O_4$
Mol. Wt.: 160.2

pinolenic acid
a C_{18} non-methylene-interrupted triene acid 5c9c12c-*octadecatrienoic acid* present in tall oil and in many conifer seed oils where it may be accompanied by its C_{20} homologue and by similar C_{18} and C_{20} tetraene acids. See also *columbinic acid.*

pistachio oil
an oleic-rich oil (ca 70%) from pistachio nuts (*Pistacia vera*). See also *speciality oils.*

planar chromatography
see *thin-layer chromatography.*

plaques
formed in an advanced stage of *atherosclerosis* by deposition of lipids and connective tissue components.

plasma lipids
see *lipoproteins*.

plasmalogens
lipids containing α-unsaturated ethers (alkenylglycerols) in position *sn*–1. These are hydrolysed under acidic conditions to give aldehydes [R^{1}CH$_2$CHO from the structure below]. Plasmalogens occurs in animal tissues as 1-alk-1'-enyl-2-acyl-3-phosphocholine-*sn*-glycerols and, mainly, 1-alk-1'-enyl-2-acyl-3-phosphoethanolamine-*sn*-glycerols. Their biochemical significance is not understood.
E. Klenk, Plasmalogens, Progress in the Chemistry of fats and Other Lipids, 1963, 6, 1-30

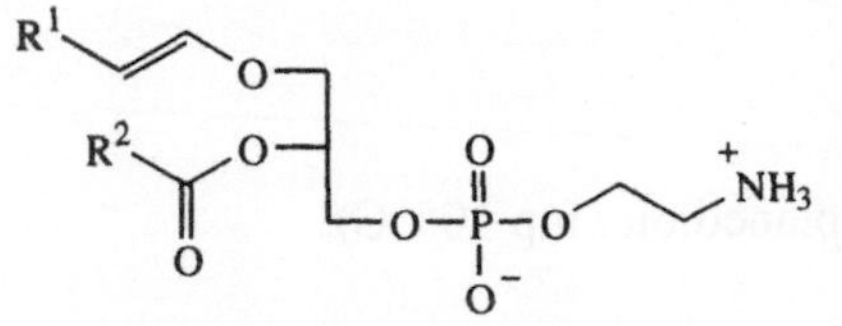

Typical plasmalogen
1-alk-1'-enyl-2-acyl-3-phosphoethanolamine-*sn*-glycerol

platelet activating factor
1-*O*-alkyl-2-acetyl-*sn*-glycero-3-phosphocholine. Phospholipid molecules of this type have biological activity (aggregation, inflammatory, oedemic) at very low concentrations and are specifically bound to receptors in platelets.
R. Bittman, Lipid Synthesis and Manufacture (ed. F.D. Gunstone) Sheffield Academic Press, Sheffield (1999) pp.185-207.

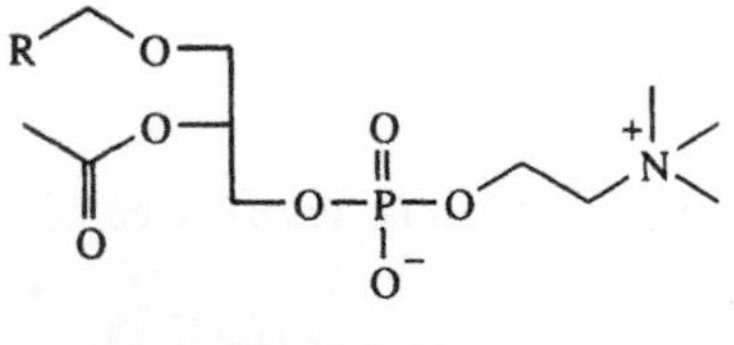

Platelet activating factor
1-*O*-alkyl-2-acetyl-*sn*-glycero-3-phosphocholine

podocarpic acid
see *eicosatrienoic acid.*

POEMS
see *sorbitan monostearate.*

polar head group
the term applied to the polar part of natural and synthetic *amphiphilic* substances.

polar lipids
lipids with polar groups as the head group e.g. *phospholipids* and *glycolipids.* See also *neutral lipids.*

polyenoic acids
acids with more than one olefinic centre. The double bonds usually have *cis* (Z) configuration and are methylene-interrupted. See individual acids such as *linoleic* and *linolenic acids*, also *non-methylene interrupted acids.*
J.-M. Vatele, Lipid Synthesis and Manufacture (ed. F.D. Gunstone) Sheffield Academic Press, Sheffield (1999) pp.1-45.

polyglycerol
oligomers of glycerol. These are polyethers with 2–10 glycerol units but mainly 2–4. Poly esters of these polyhydric molecules act as *emulsifiers* and *surfactants.*

Polyglycerol

polyglycerol esters
the products of reaction between polyglycerol (mainly 2–4 units) and fatty acids (mainly palmitic, stearic, or oleic). Reaction occurs in the presence of an alkaline catalyst and the products contain 20–40% of mono ester. They are stable in the α form and are water-dispersible. They are incorporated into bakery products. E number in Europe 475; US/FDA/CFR 172854.
N. Krog, Lipid Technologies and Applications (ed. F.D. Gunstone and F.B. Padley) Marcel Dekker, New York (1997), pp.521-534.

polymorphism

alternative crystal structures in the solid state which give complex melting behaviour with multiple melting points. Extensively studied in acids and their glycerol esters. See *crystal structure of triacylglycerols*.

P.J. Lawler and P.S. Dimick, Food Lipids: Chemistry, Nutrition, and Biotechnology, (ed. C.C. Akoh and D.B. Min) Marcel Dekker, New York (1998) pp.229-250. K. Larsson, Lipids — Molecular Organisation, Physical Functions and Technical Applications, The Oily Press, Dundee, (1994).

polyphosphoinositides

see *phosphatidylinositides*.

polyprenols

a series of isoprenoid alcohols, sometimes named terpenols, with the general structure shown. Examples of individual members of this group are geraniol (C_{10}), farnesol (C_{15}), (structure below), and geranylgeraniol (C_{20}).

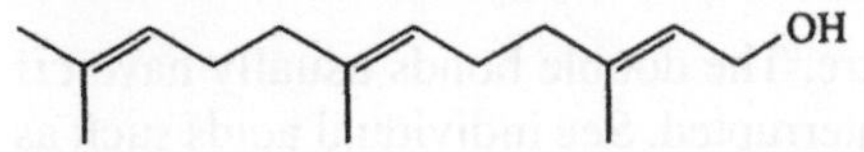

Farnesol
3,7,11-Trimethyl-2*E*,6*E*,10-dodecatrien-1-ol
$C_{15}H_{26}O$
Mol. Wt.: 222.4

pomace

see *olive oil (grades)*.

POP

1,3-dipalmitoyl-2-oleoyl glycerol. This triacylglycerol is present in *cocoa butter* (*ca.* 15%) and *confectionery fats* and is important for its physical properties (melting curves).

PORAM

Palm Oil Refiners' Association of Malaysia representing the interest of refiners in the *oil palm* industry.

PORIM

Palm Oil Research Institute of Malaysia. A research organisation devoted to the botany, chemistry and technology, and techno-economics of the *oil palm* and its products.

Yusof Basiron, Palm Oil Research Institute of Malaysia (PORIM), Lipid Technology Newsletter, 2000, 6, 19-21.

PORLA

Palm Oil Registration and Licensing Authority. A government agency charged with overseeing and monitoring the orderly development of *palm oil* in Malaysia.

POSt

1-palmitoyl-2-oleoyl-3-stearoyl glycerol. This triacylglycerol is present in *cocoa butter* (*ca.* 40%) and *confectionery fats* and is important for its physical properties (melting curves).

preen waxes

see *uropygial glands.*

pre-gastric lipases

see *lipases.*

pressing

after size reduction (if necessary), cooking at 90–115°C, and flaking, seeds are put through a screw press at 4.5 kg/mm² to squeeze out the oil. The residual protein meal may still contain 3–6% oil. Lower pressures are used if the seed meal is to be subsequently solvent extracted. Improved yields are obtained if the seed is first treated with a suitable enzyme to break down the carbohydrate in the seed coat.

J.-M. Fils, Edible Oil Processing (eds W. Hamm and R.J. Hamilton), Sheffield Academic Press, Sheffield (2000), pp.47-78.

pristanic acid

2,6,10,14-tetramethylpentadecanoic acid. A branched chain C_{19} acid derived from *phytanic acid* by *α-oxidation*. It is present in many land and marine animal fats at low levels.

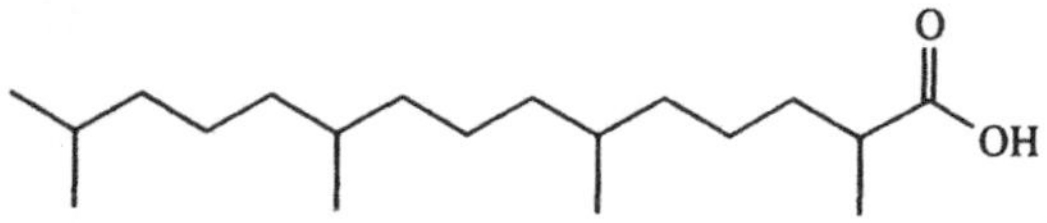

Pristanic acid
2,6,10,14-Tetramethylpentadecanoic acid
$C_{19}H_{38}O_2$
Mol. Wt.: 298.5

Progress in Lipid Research
a review series started in 1952 under the title `Progress in the Chemistry of
Fats and Other Lipids'. Changed to the above title in 1980. Published by
Pergamon Press (now Elsevier).

pro-oxidants
materials which promote oxidation. These include pre-formed *hydroperoxides*
and metals (particularly copper and iron) all of which promote the
chain-initiating step of *autoxidation*.
E.N. Frankel, Lipid Oxidation, The Oily Press, Dundee (1998).

propanoic acid
the C_3 alkanoic acid. Associated with appropriate enzymes it can act as a
precursor for odd and for branched-chain acids.

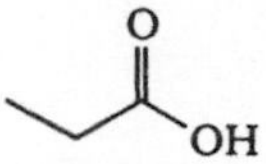

Propanoic acid
(3:0)
$C_3H_6O_2$
Mol. Wt.: 74.1

propylene glycol esters
esterified with palmitic or stearic acid in the presence of an alkaline catalyst,
propylene glycol ($CH_3CHOHCH_2OH$) gives a mixture of mono (~55%) and
di (~45%) acyl compounds. This mixture can be distilled to give 90%
concentrates of the monoacyl derivative with mp about 40°C. These
are effective emulsifiers in whipped emulsions (toppings),
bakery shortenings, cake mixes, etc. The monostearate has E number 477 and
US/FDA/CFR 172,856.

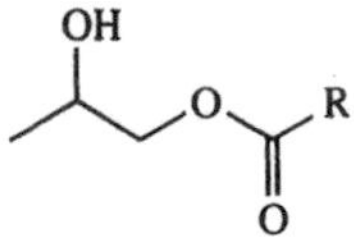

Propylene glycol monoester
N. Krog, Lipid Technologies and Applications (ed. F.D. Gunstone and F.B. Padley) Marcel Dekker,
New York (1997), pp.521-534.

propyl gallate

esters of gallic acid can be used as *antioxidants* of which the propyl ester (E310) is most commonly employed. It is less soluble than *butylated hydroxy anisole* or *butylated hydroxy toluene* and does not generally survive cooking. It is effective when used with butylated hydroxy anisole and can be used at levels up to 100 ppm.

E.N. Frankel, Lipid Oxidation, The Oily Press, Dundee (1998)

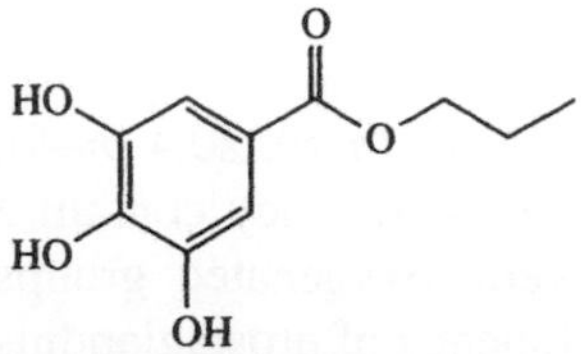

Propyl gallate
Propyl-3,4,5-trihydroxybenzoate
$C_{10}H_{12}O_5$
Mol. Wt.: 212.2

prostacyclin

PGI$_2$, a metabolite of arachidonic acid (*eicosatetraenoic acid*) formed by enzymic oxidation and cyclisation. Prostacyclin acts as a muscle relaxant and inhibits blood platelet aggregation. See also *prostaglandins, leukotrienes, eicosanoid cascade.*

N.H. Wilson, Lipid Synthesis and Manufacture (ed. F.D. Gunstone) Sheffield Academic Press, Sheffield (1999) pp.127-161.

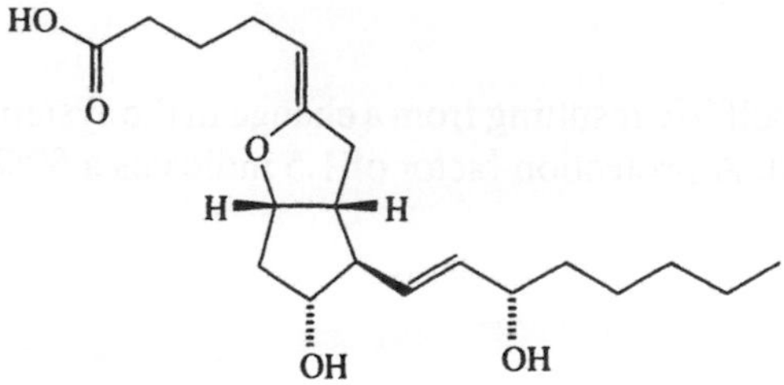

Prostacyclin
6,9α-Epoxy-11α.15S-dihydroxyprosta-5Z,13E-dien-1-oic acid
$C_{20}H_{32}O_5$
Mol. Wt.: 352.5

prostaglandin H synthase
an enzyme catalysing the first step in the synthesis of *prostaglandins*, *thromboxanes* and *prostacyclin*. It promotes the conversion of arachidonic acid (*eicosatetraenoic acid*), first to a *hydroperoxide* (prostaglandin G_2) and then to a hydroxy acid (prostaglandin H_2). It combines both cyclo-oxygenase and peroxidase activity.
N.H. Wilson, Lipid Synthesis and Manufacture (ed. F.D. Gunstone) Sheffield Academic Press, Sheffield (1999) pp.127-161.

prostaglandins
a series of oxidised metabolites of C_{20} polyene acids [20:3 (*n*–6), 20:4 (*n*–6), 20:5 (*n*–3)] especially arachidonic (*eicosatetraenoic acid*). They contain a cyclopentane system linking C8 to C12 and several oxygenated groups (peroxy, hydroxy, oxo). An absolute or relative deficiency of prostaglandins has been demonstrated in many diseased and clinical conditions. Excess can cause inflammation. Bioactive prostaglandins are formed rapidly from their precursors and are also rapidly degraded to less active metabolites. See also *prostanoids, leukotrienes, thromboxanes*.
N.H. Wilson, Lipid Synthesis and Manufacture (ed. F.D. Gunstone) Sheffield Academic Press, Sheffield (1999) pp.127-161.

prostanoids
metabolites of unsaturated C_{20} acids, especially arachidonic (*eicosatetraenoic acid*), including *prostaglandins, leukotrienes* and *thromboxanes*. These are products of the *eicosanoid cascade*.
D.M. Klurfeld, Food Lipids: Chemistry, Nutrition, and Biotechnology, (ed. C.C. Akoh and D.B. Min) Marcel Dekker, New York (1998) pp.495-505. N.H. Wilson, Lipid Synthesis and Manufacture (ed F.D. Gunstone) Sheffield Academic Press, Sheffield (1999) pp.127-161.

protection factors
a figure indicating the effect on shelf life resulting from a change in the system such as addition of an antioxidant. A protection factor of 1.5 indicates a 50% increase in shelf life.

PS
see *phosphatidylserines*.

pseudoeleostearic acid
trivial name for a conjugated trienoic C_{18} acid (10*t*12*t*14*t*-octadecatrienoic acid) produced from linolenic acid by alkali-isomerisation.

Pseudomonas fluorescens
see *lipases*.

psychosine
trivial name of a deacylated *galactosylceramide*: *O*-sphingosylgalactoside.

psyllic acid
trivial name for tritriacontanoic acid (33:0).

PUFA
polyunsaturated fatty acids. See *polyenoic acids*.

pulmonary surfactant
complex lipoprotein material absorbed in the alveolar air liquid interface of the lungs. It lowers surface tension and hence reduces the work of lung expansion. The lipid component (approximately 90%) is mainly *phosphatidylcholines* (70–80%) of which dipalmitoyl-phosphatidylcholine (DPPC) is a dominating species.

J.L. Harwood, The Lipid Handbook (2nd edition) (F.D. Gunstone, J.L. Harwood, F.B. Padley, ed.) Chapman and Hall, London, 1994, pp.689-693. J.J. Batenburg et al., The Lipids of Pulmonary Surfactant: Dynamics and Interactions with Proteins, Progress in Lipid Research, 1998, 37, 235-276.

punicic acid
a C_{18} acid with conjugated triene unsaturation (9c11t13c-octadecatrienoic acid) present in pomegranate seed oil. See also *conjugated unsaturation*.

pyrulic acid
an acetylenic C_{17} acid (8a,10t-17:2) present in the seed oil from *Pyrularia pubera*.

Pyrulic acid
10*E*-Heptadecen-8-ynoic acid
$C_{17}H_{28}O_2$
Mol. Wt.: 264.4

R

rancidity

a fat/oil is described as rancid on the basis of its undesirable odour and flavour resulting from oxidative deterioration (aldehydes and other short-chain compounds) or from hydrolysis (short-chain and medium-chain acids). See also *peroxide value, anisidine value, totox value, induction period*.

F. Shahidi and U.N. Wanasundara, Food Lipids: Chemistry, Nutrition, and Biotechnology, (ed. C.C. Akoh and D.B. Min) Marcel Dekker, New York (1998) pp.377-396.   E.N. Frankel, Lipid Oxidation, The Oily Press, Dundee (1998).

Rancimat

a machine for the measurement of resistance to oxidative *rancidity*. The method is based on changes in the conductivity of the steam distillate from material being oxidised at 100–120°C. These changes result mainly from short-chain acids (C_1–C_3) which are tertiary products of fat oxidation. Results obtained at these elevated temperatures must be interpreted with care when using them to predict shelf life since the mechanisms of oxidation change with temperature.

E.N. Frankel, Lipid Oxidation, The Oily Press, Dundee (1998).

randomisation

any process by which the fatty acids in triacylglycerols, which are not normally randomly distributed in natural lipid mixtures, are changed to a random distribution pattern. This has consequences for some physical (e.g. melting behaviour) and nutritional properties. Such a change occurs in interesterification.

A. Rosendaal and A. R. Macrae, Lipid Technologies and Applications (ed. F.D. Gunstone and F.B. Padley) Marcel Dekker, New York (1997), pp.223-263.

rapeseed oil

formerly known as colza oil, the seed oil of *Brassica napus* or *B. campestris*. Typically this oil was rich in *erucic acid*, which is still available from high-erucic rapeseed oil (*HEAR*) or from *crambe oil*. The variety low in erucic acid (<5% or <2%) and also in *glucosinolates* (*LEAR*, double zero) is now more important. At about 12.6 million tonnes a year, it is the third largest source of oil after *soya* and *palm* and is produced mainly in Western Europe, China, India, and Canada. In this last country it is known as *canola oil*. The oil typically contains palmitic (4), stearic (2), oleic (56), linoleic (26), and linolenic acid (10%) and in one example its major triacylglycerols included

LLL (5), LLO and LnOO (19), LOO (27), and OOO (41%). Rapeseed lends itself to *genetic manipulation* and rapeseed oil containing a lower level of linolenic acid or higher levels of lauric, stearic, or oleic acid or new acids such as δ-linolenic, ricinolenic, or vernolic are being developed for commercial exploitation. See also *lauric-canola*.

ratite
big flightless birds such as emu, ostrich, rhea, kiwi and cassawary. There is a growing interest in the oils obtained from these birds for use in cosmetics. It has been reported that they show anti-arthritic and anti-inflammatory properties.

M.C. Craig-Schmidt, Ratite oils — composition and claimed beneficial effects, Lipid Technology Newsletter, 1999, 5, 80-83

RBD
used to describe oil that has been *refined, bleached* and *deodorized*.

red palm oil
a palm oil containing most (>80%, 500–700 ppm) of the carotenes present in the crude oil. These are normally much reduced in RBD oil. After pre-treatment to remove impurities and phospholipids the oil is *deodorized* and *deacidified* by molecular distillation at low temperatures and reduced pressures. This carotene-rich oil has greater nutritional value than the RBD oil.

refining
the combination of processes by which a crude oil can be converted to a bland oil with low levels of free acid, phospholipids, trace metals, pigments, oxidation products etc. There are two common sequences: (i) *degumming* followed by *neutralization, bleaching* and *deodorization* and (ii) *bleaching* followed by *physical refining*. During refining some valuable minor components such as *tocols, phytosterols* and *carotenes* are removed and it may be desirable to add them back to the oil to increase its oxidative stability and its dietary value. Some of these (*lecithins, tocopherols, sterols*) can be recovered from the *deodorizer distillate*.

D.A. Allen, Lipid Technologies and Applications (ed. F.D. Gunstone and F.B. Padley) Marcel Dekker, New York (1997), pp.137-167. L.A. Johnson, Food Lipids: Chemistry, Nutrition, and Biotechnology, (ed. C.C. Akoh and D.B. Min) Marcel Dekker, New York (1998) pp.181-228. W. De Greyt and M. Kellens, Edible Oil Processing (eds W. Hamm and R.J. Hamilton), Sheffield Academic Press, Sheffield (2000), pp.79-128. J.P. Clark, Tocopherols and sterols from soybeans, Lipid Technology, 1996, 8, 111-114.

Refsum's disease

a rare human metabolic disorder, characterized by an accumulation of *phytanic acid* (from dietary *phytols*), probably due to a block in the α-*oxidation* system.

regiospecific analysis

analytical procedures (chemical or physical) which distinguish fatty acid in the α (*sn*-1 and 3) and β (*sn*-2) positions in glycerol esters but do not distinguish between the two α positions. See also *stereospecific analysis*.

rendering

a method of obtaining fat from animal tissue involving hashing (reduction to small pieces) followed by steaming either with wet steam (~100°C) followed by decanting the oil and centrifuging or with dry steam followed by pressing and separation of the oil/water mixture with a decanter and centrifuge.

retinol

vitamin A. Formed in animals from its precursor β-carotene. Necessary for the vision system. See also *carotenoids*.

Retinol
3,7-Dimethyl-9-(2,6,6trimethyl-1-cyclohexen-1-yl)-2*E*,4*E*,6*E*,8*E*-nonatetraen-1-ol
$C_{20}H_{30}O$
Mol. Wt.: 286.5

reversed micelles

also called microemulsions; water aggregates in an oil phase. See also *micelles*.

reversed-phase chromatography

see *high-performance liquid chromatography, thin-layer chromatography*.

reversion

the development of undesirable off-flavours especially in oil which has previously been refined. It is particularly noticeable in oils which contain α-linolenic acid such as soybean.

E.N. Frankel, Lipid Oxidation, The Oily Press, Dundee (1998).

Revue Française des Corps Gras
journal published by the French Society for Lipid Research (Association Française pour l'Étude des Corps Gras) until 1993. Now included in *Oleagineux Corps gras Lipides.*

retro fats
in contrast to fats which are esters of medium-chain and long-chain acids with glycerol, retro fats are esters of long-chain alcohols with small polycarboxylic acids such as citric. They are resistant to lipolysis and therefore are not absorbed from the intestines and have zero energy value.

rhamnolipids
glycolipids containing the sugar rhamnose and 3-hydroxy carboxylic acids. These lipids are isolated from microorganisms (e.g. *Pseudumonas aeruginosa*).

Rhizopus lipase

a commercially available lipase from the mould *Rhizopus arrhizus* with a specificity for the *sn*–1 and *sn*–3 positions in triacylglycerols. See *lipase, lipolysis.*

Rhizopus species
see *lipases.*

rice bran oil
a by-product of the manufacture of white rice. Production is estimated to be about 0.45 million tonnes a year. The oil is an excellent salad and frying oil with high oxidative stability. The major fatty acids are palmitic (12–18, typically 16%), oleic (40–50, typically 42%), and linoleic (29–42, typically 37%). The oil is reported to lower serum *cholesterol* levels by virtue of the oryzanols (1.5–2.0% of the oil). These are part of the *unsaponifiable matter* and are *ferulic acid* esters of *sterols* and triterpene alcohols. It may be added to other oils to enhance their oxidative stability, see *Good-Fry oil.*

B. Sayre and R. Saunders, Rice bran and rice bran oil, Lipid Technology, 1990, 2, 72-76.
F.T. Orthoefer, Bailey's Industrial Oil and Fat Products, (ed. Y.H. Hui) John Wiley & Sons, New York (1996), Volume 2, pp.393-409.

ricin (huile de)
French name for *castor oil.*

ricinelaidic acid
the *trans* isomer of *ricinoleic acid.*

ricinoleic acid
the most common of all hydroxy fatty acids comprising about 90% of the acids of *castor oil.* It is 12(*R*)-hydroxy-9-*cis*-octadecenoioc acid and is the source of many compounds used on an industrial scale. See also: *hydroxy acids, sebacic acid, undecenoic acid, octadecadienoic acids,* and *hydroxystearic acid.*
L. Rebrovic and F.D. Gunstone, Oxidative cleavage of unsaturated fatty acids, Lipid Technology, 1996, 8, 135-137. M. Schwitzer, Oleochemicals from ricinoleic acid, Lipid Technology, 1991, 3, 117-121.

Ricinoleic acid
12*R*-Hydroxy-9Z-octadecenoic acid
$C_{18}H_{34}O_3$
Mol. Wt.: 298.5

Ricinus communis
see *castor oil.*

Rilsan™
a nylon polymer made from 11-aminoundecanoic acid, itself obtained from *ricinoleic acid* via *10-undecenoic acid.*

(La) Rivista Italiana delle Sostanze Grasse
journal published by the Italian Society for Lipid Research (Società Italiana per lo Studio delle Sostanze Grasse).

riz (huile de)
French name for *rice bran* oil.

RME
rape methyl esters. Made from rapeseed oil, especially in Europe, for use as *biodiesel.*

rosilic acid
10-hydroxyoctadecanoic acid present in some leaf waxes.

rumen biohydrogenation
ruminant animals partially hydrogenate dietary unsaturated lipids through the activity of microorganisms present in the rumen. This destroys *essential fatty acids* by converting them to *conjugated linoleic acid, trans* monoenes, and saturated fatty acids. See *rumenic acid* and *vaccenic acid.*

rumenic acid
trivial name for *9c11t-octadecadienoic acid,* the major *conjugated linoleic acid* formed by *rumen biohydrogenation* of *linoleic acid* and an intermediate in the production of *vaccenic acid.*

S

sabinic acid

12-hydroxydodecanoic acid, present in *Juniperus oxycedrus* leaves (mp 78–79°C, 84–85°C).

Sabinic acid
12-Hydroxydodecanoic acid
$C_{12}H_{24}O_3$
Mol. Wt.: 216.3

safflower oil

a minor seed oil from the plant *Carthamus tinctorius*, grown particularly in India. Normally it is a linoleic-rich oil (~75% linoleic acid) with LLL (47%), LLO (19%), and LLS (18%) as the major triacylglycerols. A safflower oil rich in oleic acid (~74%) has also been developed (saffola).

J. Smith, Bailey's Industrial Oil and Fat Products, (ed. Y.H. Hui) John Wiley & Sons, New York (1996), Volume 2, pp.411-455. J.R. Smith, Safflower, AOCS Press, Champaign, Illinois, 1996.

saffola

see *safflower oil.*

salad oils

oils, usually refined and sometimes fractionated, used with salad. The oils should remain liquid even when held at 0°C.

T.L. Mounts, Lipid Technologies and Applications (ed. F.D. Gunstone and F.B. Padley) Marcel Dekker, New York (1997), pp.433-451. R.G. Krishnamurti and V.C. Witte, Bailey's Industrial Oil and Fat Products, (ed. Y.H. Hui) John Wiley & Sons, New York (1996), Volume 3, pp.193-223.

Salatrim™

the name is derived from *s*hort and *l*ong chain *a*cid *tri*glyceride *m*olecule. Salatrim is made by *interesterification* of short-chain triacylglycerols (*triacetin* and/or *tripropionin* and/or *tributyrin*) with a fully hydrogenated vegetable oil (*soybean, canola, cottonseed*). The product has reduced energy value (ca 5 kcal/g) and is designed for use in chocolates, dairy products and salty snacks.

J. W. Finley, A. McDonald et al., Lipid Technologies and Applications (ed. F.D. Gunstone and

F.B. Padley) Marcel Dekker, New York (1997), pp.501-520. M.H. Auerbach, P.W. Chang, et al., Salatrim reduced-calorie triacylglycerols, Lipid Technology, 1997,9,137-140.

sal fat
fat from the seeds of *Shorea robusta* which is rich in stearic and oleic acids (each 40–45%) and has a melting behaviour like *cocoa butter*.

sand eel
an industrial fish giving both oil and meal, see *fish oils*.

santalbic acid
see *ximenynic acid*.

sap. equiv.
see *saponification equivalent*.

saponification
literally soap-making. This name is usually applied to the alkaline hydrolysis of oils and fats yielding glycerol and the sodium or potassium salts of the long-chain acids.

saponification equivalent
the quantitative reaction of an oil or fat with alkali provides a measure of the average molecular weight of the acyl chains. The results can be expressed as saponification equivalent (the amount of oil or fat saponified by one gram mole of potassium hydroxide) or saponification value (the number of mg of potassium hydroxide required to saponify one gram of oil or fat), $SE = 56108/SV$. See also *Fourier-Transform infrared*. Standard methods are described by AOCS [Cd 3 25(89)] and IUPAC (2,202 and 3,122).

saponification value
see *saponification equivalent* and *Fourier-transform infrared*.

sap. value
see *saponification equivalent*.

sardine oil
sometimes called pilchard oil. A commercial fish oil varying in composition with geographical location and season of catching. Contains variable amounts of *eicosapentaenoic acid* (EPA) and *docosahexaenoic acid* (DHA).

sativic acid
9,10,12,13-tetrahydroxyoctadecanoic acid. This acid exists in many stereochemical forms. Oxidation of linoleic acid with dilute alkaline permanganate (a *cis* addition process) gives sativic acid.

saturated acids
fatty acids without carbon–carbon unsaturation. For example: *lauric* (12:0), *myristic* (14:0), *palmitic* (16:0) and *stearic* (18:0) acids. See also *alkanoic acids.·*

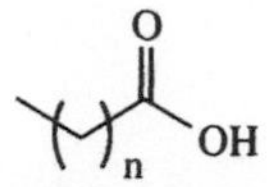

Saturated acids

sciadonic acid
trivial name of 5*c*11*c*14*c-eicosatrienoic acid.*

scoliodonic acid
a 24:5 acid reported to be present in fish oils but not fully identified.

sebacic acid
the C_{10} dibasic acid (decanedioic, mp 134°C) produced from *ricinoleic acid* by alkali fusion. Used in the production of polyesters and polyamides.

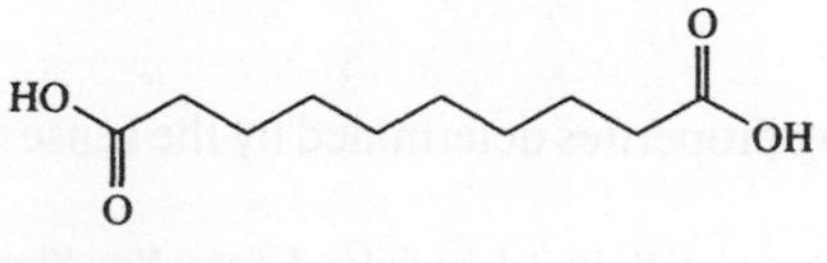

Sebacic acid
Decanedioic acid
$C_{10}H_{18}O_4$
Mol. Wt.: 202.2

sebum
a mixture of non-polar lipids produced by sebaceous glands in the skin and secreted onto the skin surface. The human sebum is composed of triacylglycerols, wax esters and squalene and is produced in association with hair follicles. See also skin lipids.

secondary oxidation products
usually refer to the products of decomposition of lipid hydroperoxides. They
are responsible for the development of off-flavours even at very low
concentrations. See *anisidine value*.

selachyl alcohol
sn–1 glycerol ether of oleyl alcohol, present in diacyl form in several fish
liver oils.

Selachyl alcohol
2*S*,3-Dihydroxy-1-octadec-9'Z-enyloxypropane
$C_{21}H_{42}O_3$
Mol. Wt.: 342.6

seminolipid
the main glycolipid, which is a complex sulpholipid, in testes and sperm.

sensory evaluation vocabulary
an agreed collection of words used by tasting panels to describe the sensory
testing of food products.
*K. Warner, Bailey's Industrial Oil and Fat Products, (ed. Y.H. Hui) John Wiley & Sons, New York
(1996), Volume 1, pp.105-144.*

sensory test
evaluation of food products by measuring properties determined by the senses
(sight, smell, hearing, touch and taste).
*K. Warner, Bailey's Industrial Oil and Fat Products, (ed. Y.H. Hui) John Wiley & Sons, New York
(1996), Volume 1, pp.105-144.*

sequestrant
see *chelator*.

sesame oil
the seed oil of the plant *Sesamum indicum* grown mainly in India and China
but also in Myanmar (Burma), Sudan, and Mexico. The annual production is
~0.7z million tonnes. The seed has 40–60% oil with almost equal levels of
oleic (range 35–54, average 46%) and linoleic acid (range 39–59, average

46%). The oil contains sesamin (0.5–1.1%) and sesamolin (0.3–0.6%) and has high oxidative stability due to the presence of natural *antioxidants*. It may be added to other oils to enhance their oxidative stability See *Good-Fry oil*.

S.S. Deshpande, U.S. Deshpande, et al., Bailey's Industrial Oil and Fat Products, (ed. Y.H. Hui) John Wiley & Sons, New York (1996), Volume 2, pp.457-495.

Sesamum indicum (Pedaliceae)
see *sesame oil*.

SFC
see *solid fat content*.

SFI
see *solid fat index*.

shark liver oil
a marine oil containing diacylated glycerol ethers (*batyl, chimyl, selachyl alcohols*) and *squalene*. See *ether lipids*.

shea butter
obtained from the kernels of *Butyrospermum parkii*. The oil is rich in stearic and oleic acids (each about 45%) and provides a *cocoa butter equivalent* after fractionation.

shea/karite/*Butyrospermum*
This fat has a high level of unsaponifiable material (5–7%) which is mainly triterpenes. It is used in cosmetics especially for suntan lotions. See *Butyrospermum parkii*.

shellac
the hardened viscous secretion (lac) of an insect parasite on certain trees in India, Myanmar (Burma), and Thailand. It is a polyester the hydrolysis products of which include some long-chain hydroxy acids (*aleuritic acid, butolic acid*).

shibic acid
a 26:5 acid reported to be present in fish oils but not fully identified.

Shorea robusta
the seed fat (sal) of this Indian tree is rich in stearic (33–49%) and oleic

(33–48%) acids and contains 65–70% of StOSt among its glycerol esters. The value rises to over 90% in sal stearin. It can be used as a *cocoa butter equivalent.*

Shorea stenoptera

the seed fat (Illipe, Borneo tallow) of this tree, grown in Malaysia and Indonesia, is rich in palmitic (~20%), stearic (~40%) and oleic (~40%) acids. It contains 80–90% of *SOS* in its glycerol esters and can be used as a *cocoa butter equivalent.*

shortening

a general description of a solid fat used in baking and cooking. Originally the term was applied to fats producing short texture (*i.e.* easily broken) in baked goods.

E. Flack, Lipid Technologies and Applications (ed. F.D. Gunstone and F.B. Padley) Marcel Dekker, New York (1997), pp.305-327. D.J. Metzroth, Bailey's Industrial Oil and Fat Products, (ed. Y.H. Hui) John Wiley & Sons, New York (1996), Volume 3, pp.115-160. R.D. O'Brien, Bailey's Industrial Oil and Fat Products, (ed. Y.H. Hui) John Wiley & Sons, New York (1996), Volume 3, pp.161-192.

sialic acid

trivial name for N-acetylneuraminic acid (NANA). A constituent of *gangliosides.*

Sialic acid
5-(Acetylamino)-3,5-dideoxy-D-glycero-D-galacto-2-nonulosonic acid
$C_{11}H_{19}NO_9$
Mol. Wt.: 309.3

silica

SiO_2; sometimes termed silicic acid or silica gel. See *adsorption chromatography* and *bleaching.*

silica chromatography

see *adsorption chromatography.*

silver ion chromatography

argentation chromatography. The interaction of silver ions with double bonds is exploited in several modes of chromatography (e.g. thin-layer chromatography and high-performance liquid chromatography). It can be used to separate *cis* and *trans* species and saturated/unsaturated acids/esters according to the number of *(cis)* double bonds present. These include intact lipids such as wax esters and triacylglycerols.

B. Nikolova-Damyanova, *Advances in Lipid Methodology — One (ed. W.W. Christie) The Oily Press, Dundee, (1992) pp.181-237. G. Dobson, Review: Silver ion chromatography of lipids and fatty acids, Journal of Chromatography B, 671, 1995, 197-222*

Simmondsia chinensis

see *jojoba oil.*

simple lipids

lipids which on hydrolysis give no more than two primary products, e.g. triacylglycerols, cholesterol esters. These are sometimes termed neutral lipids. See also *complex lipids.*

sitosterol

an abundant plant sterol (mp 137°C).

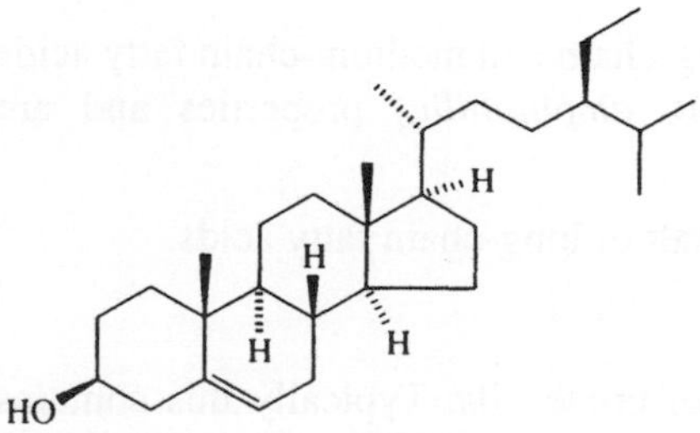

β-Sitosterol
3β-Stigmast-5-en-3-ol
$C_{29}H_{50}O$
Mol. Wt.: 414.7

skin lipids

normally referring to the lipid matrix surrounding the cells in strateum corneum which provides a permeability barrier for the skin. These lipids consist mainly of *ceramides*, *cholesterol* and *fatty acids*. See also sebum.

K. Larsson, Lipids — Molecular Organization, Physical Functions and Technical Applications, The Oil Press, Dundee (1994), pp.147-154. P.W. Wertz, Phospholipids: Characterization, Metabolism, and Novel Biological Applications (eds G. Cevc and F. Paltauf), AOCS Press, Champaign, Illinois (1995), pp.139-158. V.A. Ziboh, Metabolism and function of skin lipids, Progress in Lipid Research, 1988, 27, 81-105.

slip (melting) point

the *softening point* of oils in their natural state. The test is not applicable to samples which have been melted or otherwise modified. See also *melting point*.

small unilamellar vesicles

SUV. See *liposomes*.

smoke point

the temperature at which oils and fats begin to produce smoke.

SMS

see *sorbitan monostearate*.

sn

see *stereospecific numbering*.

soap

(i) the sodium or potassium salts of long-chain and medium-chain fatty acids which have surface-active (*detergent, amphiphilic*) properties and are commonly used for cleaning.

(ii) the general name for any metallic salt of long-chain fatty acids.

soapstock

a by-product from the neutralisation of crude oils. Typically this contains water (45%), free fatty acid (10%), glycerol esters (12%), and phospholipids and other compounds (24%). It is used as an animal feed and as a source of fatty acids. Attempts are being made to develop high value uses.

Societa Italiana per lo Studio della Sostanze Grasse

the Italian Society for Lipid Research. Publishes *La Rivista Italiana della Sostanze Grasse*.

sodium stearoyl lactate

the product of reaction of fatty acids with lactic acid in the presence of sodium (or calcium) hydroxide. The products (mp ~45°C) are water-dispersible, anionic emulsifiers. They act as dough strengtheners and as starch-complexing agents in bakery products.

N. Krog, Lipid Technologies and Applications (ed. F.D. Gunstone and F.B. Padley) Marcel Dekker, New York (1997), pp.521-534.

Sodium stearoyl lactate
$C_{21}H_{39}NaO_4$
Mol. Wt.: 378.5

softening point
the temperature at which a fat becomes sufficiently fluid to slip or run.

solid fat content
the percentage of solid determined by pulse NMR (*nuclear magnetic resonance*) is the ratio of the response from the hydrogen nuclei in the solid phase and that from all the hydrogen nuclei in the sample. This method is now used increasingly in place of the measurement of SFC by *dilatometry*. Standard methods are described by AOCS [Cd 10 57(89)].

solid fat index
a comparison of the *dilatation* of an oil or fat with that of a fat which is completely solid. This figure should not be equated with the *solid fat content* measured by *nuclear magnetic resonance*.

solin
the generic name for flaxseed with <5% of linolenic acid in its seed oil (*Linola* is a trade mark for a type of solin). It must have a yellow seed coat to distinguish it from conventional flaxseed which furnishes *linseed oil*. It is grown in Australia, Canada, and Europe.

A.G. Green and J.C.P. Dribnenki, Linola — a new premium polyunsaturated oil, Lipid Technology, 1994, 6, 29-33.

solvent extraction
seeds, sometimes reduced in size, or cake remaining after *pressing*, are extracted with industrial hexane (a mixture of hexane and methylpentane). Solvent is recovered for re-use but some is lost in the oil (~ 2 litres/tonne) and in the meal (~ 8–10 litres/tonne).

M.A. Williams, Lipid Technologies and Applications (ed. F.D. Gunstone and F.B. Padley) Marcel Dekker, New York (1997), pp.113-135. J.-M. Fils, Edible Oil Processing (eds W. Hamm and R.J. Hamilton), Sheffield Academic Press, Sheffield (2000), pp.47-78.

sophorosides

yeast glycolipids made up of monosaccharides or disaccharides glycosidically linked to 17-hydroxyoctadecanoic and 17-hydroxyoctadecenoic acids.

Sorbestrin™

ester made from *sorbitol* and long-chain acids. These are thermally stable oils suitable as replacements for natural vegetable oils in salad dressing, baked goods and fried food.

J.W. Finley, A. McDonald, et al., Lipid Technologies and Applications (ed. F.D. Gunstone and F.B. Padley) Marcel Dekker, New York (1997), pp.501-520.

sorbic acid

sorbic acid (hexadienoic) and its sodium, calcium and potassium salts are used as permitted preservatives in margarine, usually at a level of 0.4–0.8 g/kg (though up to 2 g/kg is permitted). It occurs naturally in the lipids of aphids.

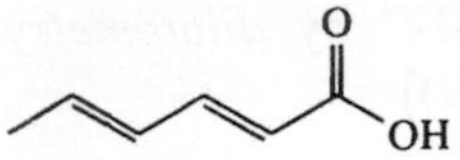

Sorbic acid
2E,4E-Hexadienoic acid
$C_6H_8O_2$
Mol. Wt.: 112.1

sorbitan

a tetrahydrofuran derivative derived from sorbitol by dehydration

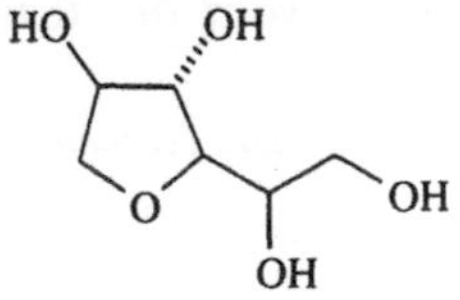

Sorbitan
2-(1,2-Dihydroxy-ethyl)-tetrahydrofuran-3,4-diol
$C_6H_{12}O_5$
Mol. Wt.: 164.2

sorbitan esters
sorbitol is a hexahydric alcohol which is readily dehydrated to a mixture of tetrahydric alcohols with a 1,4-ether link. The esters, sometimes after ethoxylation, show a wide range of *HLB* values (2–17) and act as emulsion stabilisers.

G.L. Hasenheuttl, Lipid Synthesis and Manufacture (ed. F.D. Gunstone) Sheffield Academic Press, Sheffield (1999) pp.371-400.

sorbitan monostearate
the monostearate of sorbitan (mainly the 6-derivative) which is a hydrophilic water-dispersible emulsifier. It can be reacted with ethylene oxide to give poly oxyethylene sorbitan esters (POEMS) which are among the most hydrophilic water-soluble emulsifiers available. E number in Europe 491; US/FDA/CFR 172 842.

N. Krog, Lipid Technologies and Applications (ed. F.D. Gunstone and F.B. Padley) Marcel Dekker, New York (1997), pp.521-534.   G.L. Hasenheuttl, Lipid Synthesis and Manufacture (ed. F.D. Gunstone) Sheffield Academic Press, Sheffield (1999) pp.371-400.

Sorbitan monostearate
2-(3,4-Dihydroxy-tetrahydrofuran-2-yl)-2-hydroxy-ethyl octadecanoate
$C_{24}H_{46}O_6$
Mol. Wt.: 430.6

sorbitol polyester
see *Sorbestrin^TM*.

sorbitan tristearate
a fat-soluble emulsifier with molecular structure resembling that of triacylglycerols. It is stable in the α-form and is used as a crystal modifier in margarine (where it stabilises the β′-form), as a bloom inhibitor in *chocolate* and *confectionery fats*, and as an inhibitor of crystal formation in *frying oils*. E number in Europe 492. See also *sorbitan* and *sorbitan monostearate*.

N. Krog, Lipid Technologies and Applications (ed. F.D. Gunstone and F.B. Padley) Marcel Dekker, New York (1997), pp.521-534.   G.L. Hasenheuttl, Lipid Synthesis and Manufacture (ed. F.D. Gunstone) Sheffield Academic Press, Sheffield (1999) pp.371-400.

SOS

shorthand for glycerol esters having oleic acid in the *sn*-2 position and saturated acids (usually palmitic and/or stearic) in the *sn*-1 and 3 positions. If stearic acid is meant it should be designated stearic. See *StOSt*.

soya oil

another name for *soybean oil*.

soybean oil

the seed of *Glycine max* is grown mainly as a source of high-grade protein for animal feed. The seed oil contains palmitic (about 11%, range 7–14%), oleic (about 20%, range 19–30%), linoleic (about 53%, range 44–62%), and linolenic acid (about 7%, range 4–11%). Triacylglycerols in excess of 5% typically include LeLL (7), LeLO (5), LLL (15), LLO (16), LLS (13), LOO (8), LOS (12), OOS (5), and other (19%).

Soybean oil is produced in larger amount than any other traded oil (~23 million tonnes a year) and is grown particularly in the USA, followed by Brazil, Argentina, and China. The oil is used mainly for food purposes, usually after partial *hydrogenation*, as salad oil, cooking oil, frying oil, and in *margarines* and *shortenings*. Its non-food uses in the production of coatings, *dimer acids*, and *epoxidised oil* are based on its high level of unsaturation. After suitable modification it can be used as a solvent, a lubricant and as *biodiesel*. Valuable by-products recovered during refining include *lecithin*, *tocopherols* and *phytosterols*.

Attempts to modify the fatty acid composition by seed breeding or *genetic modification* are directed to reducing the level of saturated acid or linolenic acid, or increasing the content of stearic acid.

E.F. Sipes and B.F. Szuhaj, Bailey's Industrial Oil and Fat Products, (ed. Y.H. Hui) John Wiley & Sons, New York (1996), Volume 2, pp.497-601.   J.P. Clark, Tocopherols and sterols from soybeans, Lipid Technology, 1996, 8, 111-114.

soy lecithin

the commercial mixture of triacylglycerols and *phospholipids* (50–60%; mainly *phosphatidylcholines*, *phosphatidylethanolamines* and *phosphatidylinositols*) produced from crude *soybean oil* during refining (see *degumming*). A useful emulsifying agent in food and other applications and a source of purer preparations of soybean phospholipids. See *lecithin*.

speciality oils (in USA, specialty oils)

also called gourmet oils, these are produced in relatively small volumes. They are extracted and refined with extra care and used as food oils (sometimes because of their distinctive flavour) and in toiletries, cosmetics and pharmaceuticals. They include oils that are rich in *oleic acid* (*almond, apricot, avocado, hazelnut, macadamia, pecan* and *pistachio*), rich in *linoleic acid* (*grapeseed* and *passion flower*), contain moderate levels of oleic and linoleic acid (*sesame*), or contain *α-linolenic acid* in addition to oleic and linoleic (*walnut* and *wheat germ*).

J. Hancock and C. Houghton, Exotic oils, Lipid Technology, 1990, 2, 90-94.

spectroscopy
ultraviolet, infrared, nuclear magnetic resonance and *mass spectrometry* are used increasingly in the study of lipids.

spermaceti
see *sperm whale oil*.

sperm whale oil
an excellent lubricant but now a proscribed product, replaced by synthetic esters or *jojoba oil*. It contains wax esters (~76%) and triacylglycerols (~23%) and is known as spermaceti.

spherosome
a subcellular particle in plant cells with a high content of phospholipids.

sphingoid bases
long-chain bases; mainly C_{18} or C_{20} aliphatic amines, with two (or three) hydroxyl groups and often a *trans* double bond in position 4. They are constituents of *sphingolipids* with the amine group acylated with a fatty acid and the primary alcohol group usually attached to one or more sugar residues or to phosphorylcholine. See also *phytosphingosine* and *sphingosine*.

sphingolipidoses
diseases characterized by the accumulation of different types of *sphingolipids* in various organs and tissues. They result from enzymic defects in sphingolipid metabolism.

sphingolipids
the name for lipids based on *sphingoid bases*, e.g. *ceramides, gangliosides*. See also *glycosphingolipids, glycolipids, sphingomyelin*.

K.-H. Jung and R.R. Schmidt, Lipid Synthesis and Manufacture (ed. F.D. Gunstone) Sheffield Academic Press, Sheffield (1999) pp.208-249.

sphingomyelin
the phosphorylcholine ester of *N-acylsphingosines* (ceramides). These are
hydrolysed to sphingosine(s), fatty acids, phosphoric acid and choline and are
common constituents of animal membranes.

Sphingomyelin
Ceramide-1-phosphocholine

sphingomyelinase
lipase acting on sphingomyelin equivalent to *phospholipase C* on
phospholipids. See *lipases.*

sphingosine
4*t*-sphingenine; the most common sphingoid base in animal tissues.
K.-H. Jung and R.R. Schmidt, Lipid Synthesis and Manufacture (ed. F.D. Gunstone) Sheffield
Academic Press, Sheffield (1999) p.208.

Sphingosine
2S-Amino-4E-octadecene-1,3R-diol
$C_{18}H_{37}NO_2$
Mol. Wt.: 299.5

squalane
fully hydrogenated **squalene**. This branched-chain alkane $(C_{30}H_{62})$ is used in
the cosmetics industry.

squalene
a triterpene hydrocarbon $(C_{30}H_{50})$. This is an important precursor of *lanosterol*
in the biosynthesis of *sterols*. It is present at low levels in most vegetable oils,

especially in *olive oil* (0.14–0.70%), and at still higher levels in some marine oils. It can be recovered in commercial quantities from *shark liver oil*.

Squalene
2,6,10,15,19,23-Hexamethyl-2*E*,6*E*,10*E*,14*E*,18*E*,22*E*-tetracosahexaene
$C_{30}H_{50}$
Mol. Wt.: 410.7

SSL
see *sodium stearoyl lactate*.

starflower oil
see *borage oil*.

steam refining
see *physical refining*.

stearic acid
trivial name for *octadecanoic acid* (18:0).

stearidonic acid
trivial name for 6*c*9*c*12*c*15*c*-octadecatetraenoic acid.

stearin
oils and fats can be separated into two fractions. The less-soluble higher-melting fraction is called the stearin and has a higher content of saturated acids. The other fraction is called the *olein*.

R.E. Timms, *Lipid Technologies and Applications (ed. F.D. Gunstone), Marcel Dekker, New York, New York, 1997, pp.199-222.*

stearolic acid
trivial name for 9-octadecynoic acid (mp 47–48°C), the acetylenic analogue of oleic acid. It occurs only rarely in seed oils but is readily made from oleic acid by bromination and dehydrobromination. The isomeric *octadecynoic acids* have been synthesised. See also *tariric acid* and *acetylenic acids*.

steradienes
see *stigmastadiene.*

sterculic acid
a C_{19} cyclopropene acid occurring with *malvalic acid* in the seed oils of the Malvales including *kapok seed* and *cottonseed oil.* Both these cyclopropene acids inhibit the biochemical desaturation of stearic acid to oleic acid. It gives a positive reaction in the *Halphen test.* Related acids include the 17-ynoic acid (sterculynic) and 2-hydroxysterculic acid. See also *cyclopropene acids.*

Sterculic acid
9,10-Methyleneoctadec-9-enoic acid
$C_{19}H_{34}O_2$
Mol. Wt.: 294.5

sterculynic acid
the 17,18 acetylenic analogue of *sterculic acid* present in *Sterculia alata* seed oil.

stereomutation
change of configuration, term used to describe the interchange of *cis* and *trans* isomers. Double bond migration is frequently accompanied by stereomutation. This change also occurs at high temperature during *deodorization* or with reagents that convert individual stereoisomers to an equilibrium mixture of the *cis* and *trans* isomers. These include nitrogen dioxide (from nitric acid and sodium nitrite), 3-mercaptopropionic acid, 2-mercaptoethanoic acid, and 2-mercaptoethylamine. See also *cis* and *trans.*

stereospecific analysis
multi-step analytical procedures, sometimes involving enzymic reactions but not necessarily so, or use of chiral chromatography, which give the composition of acids associated with each of the three glycerol carbon atoms (*sn*–1, 2 and 3) of triacylglycerols or triacylglycerol mixtures. It is an important part of triacylglycerol analysis. See also *stereospecific numbering* and *regiospecific analysis.*
W.W. Christie, Advances in Lipid Methodology — One (ed. W.W. Christie) The Oily Press, Dundee (1992) pp.121-148.

stereospecific numbering

convention (*sn–*) used to designate the stereochemistry of glycerol-based lipids. When the glycerol moiety is drawn in the Fischer projection with the secondary hydroxyl to the left, the carbons are numbered 1,2,3 from top to bottom.

$$H_2\overset{1}{C}-O-A^1$$
$$A^2-O-\overset{2}{\underset{|}{C}}-H$$
$$H_2\overset{3}{C}-O-B$$

Stereospecific numbering
General configuration of the backbone of glycerolipids.
A^1 and A^2 = acyl or alkyl groups. B = acyl groups in triacylglycerols, otherwise a polar group

sterol esters

sterols with the hydroxyl group esterified by fatty acids.

sterol glycosides

family of compounds consisting of carbohydrate unit(s) glycosidically linked to the hydroxyl group of (normally) plant *sterols*.

sterols

essential components of all eukaryotic cells with structural function and precursors of hormones and *bile acids*. Most vegetable oils contain 1000–5000 ppm (0.1–0.5%) of total sterols (both free and acylated with long-chain acids). *Sitosterol* is the major phytosterol (50–80% of total sterols). *Cholesterol* is a zoosterol not found in plants at a significant level. The phytosterols, recovered particularly from *soybean oil*, are used to prepare other sterols for use as pharmaceuticals. Phytosterols may be added to spreads to reduce the absorption of cholesterol. See also *Benecol™* and *vegetable oils*.
S. Li and E.J. Parish, Food Lipids: Chemistry, Nutrition, and Biotechnology, (ed. C.C. Akoh and D.B. Min) Marcel Dekker, New York (1998) pp.89-114. J.P. Clark, Tocopherols and sterols from soybeans, Lipid Technology, 1996, 8, 111-114.

stigmastadiene

this 3,5-diene is present in refined oils as a consequence of the high temperature dehydration of β-sitosterol. It can be used to distinguish such oils from less refined products which have not been exposed to high temperatures.

stigmasterol

an abundant phytosterol (mp 137°C) with the structure given below. It is

present in soybean oil the mixed sterols of which are modified to produce about 70% of the world supply of corticosteroids and sex hormones.
J.P. Clark, Tocopherols and sterols from soybeans, Lipid Technology, 1996, 8, 111-114.

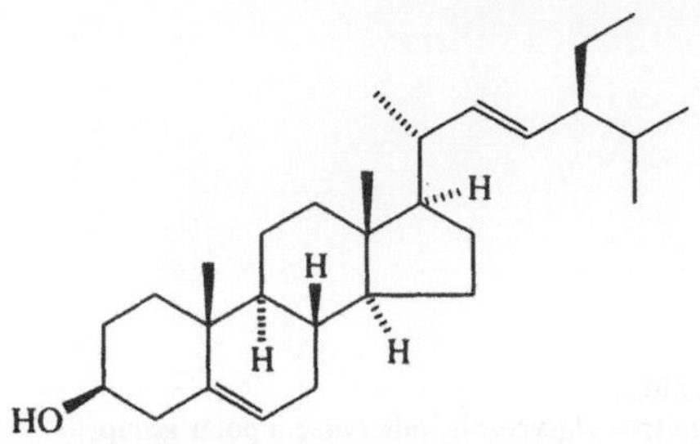

Stigmasterol
3β,22E-Stigmasta-5,22-dien-3-ol
$C_{29}H_{48}O$
Mol. Wt.: 412.7

stillingic acid

this unusual C_{10} acid (2c4t isomer) is present in *stillingia oil (Sapium sebiferum)* where it is associated with a C_8 hydroxy *allenic acid*. Together these make a C_{18} chain which appears to be confined to the *sn*–3 position. The glycerol ester, which also contains *linoleic* and *linolenic acid*, is therefore a tetra ester (an *estolide*).

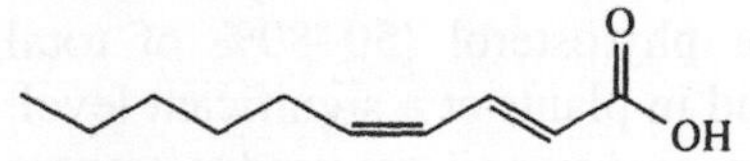

Stillingic acid
2E,4Z-Decadienoic acid
$C_{10}H_{16}O_2$
Mol. Wt.: 168.2

stillingia oil

oil from the seed of the tree *Sapium sebiferum (Stillingia sebifera)*. The outer seed coating gives a fat (*Chinese vegetable tallow*) whilst the seed gives stillingia oil. This latter contains oleic, linoleic and α-linolenic acids along with a C_8 *allenic hydroxy* acid and 2,4-decadienoic acid (*stillingic acid*).

StOSt

1,3-distearoyl-2-oleyl glycerol. An important component in *cocoa butter* where it is accompanied by other disaturated 2-oleo glycerol esters. These have a characteristic melting curve which makes cocoa butter and some other fats suitable as *confectionery fats* for the production of *chocolate*. Glycerol esters of this type can also be made enzymically from triolein and stearic acid using a 1,3-specific lipase.

straight-phase chromatography
see *high-performance liquid chromatography*.

structured fats
structured fats are designed to produce a particular property such as lower energy value, a designated melting behaviour, or specific nutritional property. Low energy fats are triacylglycerol mixtures designed to produce less than the normal 9 kcal/g when metabolised. This is achieved by incorporation of short-chain acids which have lower energy values and of long-chain acids. The latter are not completely absorbed from the intestines and metabolised and thus have apparently a lower energy value. Examples include *Salatrim^TM* and *Caprenin^TM*. Acyl derivatives of carbohydrates rather than glycerol (e.g. *Olestra^TM*) are not absorbed and whilst sharing many of the properties of *triacylglycerols* they have zero energy value. These are called fat mimetics and other examples include *Sorbestrin^TM* and *retro fats*). Other structured lipids include *Appetizer^TM* shortening, *Bohenin^TM*, *medium-chain triacylglycerols*, *Arasco^TM*, *Dhasco^TM*, enhanced fish oils and enhanced GLA-oils.

C.C. Akoh, Making new structured fats by chemical reaction and enzymatic modification, Lipid Technology, 1997, 9, 61-66. J.W. Finley, A. McDonald et al., Lipid Technologies and Applications (ed. F.D. Gunstone and F.B. Padley) Marcel Dekker, New York (1997), pp.501-520. C.C. Akoh, Food Lipids: Chemistry, Nutrition, and Biotechnology, (ed. C.C. Akoh and D.B. Min) Marcel Dekker, New York (1998) pp.699-727.

STS
see sorbitan tristearate.

suberic acid
the C_8 dibasic acid, octanedioic acid (mp 144°C).

Suberic acid
Octanedioic acid
$C_8H_{14}O_4$
Mol. Wt.: 174.2

suberin

a cork-like polymer made up of long-chain (C_{18} upwards) acids, alcohols, dibasic acids, and phenolic compounds (e.g. phellonic acid).

succinic acid

trivial name for the C_4 dibasic acid butanedioic (mp 185°C).

Succinic acid
Butanedioic acid
$C_4H_6O_4$
Mol. Wt.: 118.1

sucrose esters

a new generation of food emulsifiers made by transesterification of sucrose and methyl esters in dimethylformamide or dimethylsulphoxide solution. The products are mixtures of mono, di and tri esters. Samples containing ~70% of the mono ester are water-dispersible whilst samples which are more extensively acylated and contain only 10–30% of monoesters are oil-soluble. These compounds are used widely in Japan. E number in Europe 473; US/FDA/CFR 172859.

N. Krog, Lipid Technologies and Applications (ed. F.D. Gunstone and F.B. Padley) Marcel Dekker, New York (1997), pp.521-534. Sakidja and B.G. Swanson, Lipid Synthesis and Manufacture (ed. F.D. Gunstone) Sheffield Academic Press, Sheffield (1999) pp.347-370.

sucroglycerides

a mixture of sucrose esters (40–60%) and mono/diacylglycerols (60–40%) made by transesterification of sucrose and triacylglycerols in dimethylformamide at ~140°C. E number in Europe 474. This material is used as a food emulsifier.

Sakidja and B.G. Swanson, Lipid Synthesis and Manufacture (ed. F.D. Gunstone) Sheffield Academic Press, Sheffield (1999) pp.347-370.

sulphatides

normally refers to sulphate esters of *cerebrosides*. These are widely distributed at low levels in mammalian tissues.

sulpholipids

general term for lipids containing sulphonic acid or other sulphur-containing groups. They include the plant lipid class *sulphoquinovosyldiacylglycerols* present in plant membranes.

I. Ishizuka, Chemistry and functional distribution of sulphoglycolipids, Progress in Lipid Research, 1997, 36, 245-319.

sulphonates (α)

produced commercially from saturated acids or esters by reaction with oleum, sulphur trioxide, or chlorosulphonic acid. The products are mono acid salts or neutral salts and have a number of uses based on their surface activity and their low environmental toxicity. They biodegrade in a few days.

M.R. Porter, Lipid Technologies and Applications (ed. F.D. Gunstone and F.B. Padley) Marcel Dekker, New York (1997), pp.579-608.

$$R(CH_2)_nCH(SO_3H)COOH$$

α-Sulphonates

sulphoquinovosyldiacyl glycerols

plant sulpholipid, found in membranes of the chloroplasts especially in leaves and algae. Little is known about their metabolism.

Sulphoquinovosyldiacylglycerol

sunflower seed oil

a major vegetable oil (~9 million tonnes per annum), from the seed of *Helianthus annus* grown mainly in USSR, Argentina, Western and Eastern Europe, China, and USA. The oil normally contains 60–75% of linoleic acid and >90% of oleic and linoleic acids combined. It contains virtually no linolenic acid. Its major triacylglycerols are typically LLL (14), LLO (39), LLS (14), LOO (19), LOS (11), and other (3%). It is widely used as a salad oil, as a cooking oil, and in the production of margarine and shortenings. High-oleic varieties (*Sunola* or *Highsun, NuSun*) with about 85% and 60% oleic acid have been developed and find use as sources of oleic acid in enzymically modified products (see *Betapol*^{TM} etc). They are also used as frying oils.

H.F. Davidson, E.J. Campbell, et al., Bailey's Industrial Oil and Fat Products, (ed. Y.H. Hui) John Wiley & Sons, New York (1996), Volume 2, pp.603-689.

Sunola
the oil obtained from a high-oleic variety of *sunflower* oil containing about 85% of oleic acid. The oil has high oxidative stability and a `healthy image'. See also *NuSun*^{TM}.

supercritical fluid chromatography
chromatography based on a mobile phase in a supercritical state. Lipids may be separated on various stationary phases with carbon dioxide as the supercritical fluid to which polar solvents may be added to modify the selectivity of the separation.

P. Laakso, Advances in Lipid Methodology — One (ed. W.W. Christie) The Oily Press, Ayr (1992) pp.81-119. J.W. King et al., Supercritical fluid chromatography: A shortcut in lipid analysis, New Techniques and Applications in Lipid Analysis, (eds R.E. McDonald and M.M. Mossoba) AOCS Press, Champaign, Illinois (1997), pp.139-162. D.G. Hayes, Analysis of unusual triglycerides and lipids using supercritical fluid chromatography, New Techniques and Applications in Lipid Analysis, (eds R.E. McDonald and M.M. Mossoba) AOCS Press, Champaign, Illinois (1997), pp.163-182.

superolein
obtained by repeated *fractionation* of *palm oil*. It has a cloud point of 3–4°C and an iodine value of 64–66. The content of saturated acids is only 38% compared with values of 50 and 45 for palm oil and *palm olein*.

E. Deffense, Dry multiple fractionations; trends in products and applications, Lipid Technology, 1995, 7, 34-38.

super refining
an industrial chromatographic method of removing residual impurities from an oil that has already been *refined*. A refined product may still contain traces of undesirable components (pigments, trace metals, oxidation product, free acids, etc) which are removed by large-scale chromatography (on the multi-kilogram scale) to give highly refined products of interest to the cosmetic and pharmaceutical industries.

B. Herslöf and P. Kaufmann, Chromatographically pure and purified lipids in industrial use — a chemometric approach, Lipid Technology, 1990, 2, 100-104.

surface balance
instrument to study *monolayers* of lipids on the surface of water. See *Langmuir-Blodgett films*.

surfactants
surface-active compounds. Many long-chain compounds show amphiphilic behaviour. They contain *lipophilic* (*hydrophobic*) and *hydrophilic* (lipophobic) segments within the molecules which makes them suitable for use as *detergents, emulsifiers*, flotation agents etc.

SUV
small unilamellar vesicles. See *liposomes*.

Swift test
see *active oxygen method*.

T

TAG
see *triacylglycerols*.

tall oil
from the Swedish word (*tallolja*) for pine oil. This is a by-product of the wood pulp industry mainly in North America (~250,000 tonnes) and in Scandinavia (~90,000 tonnes). The two products differ in composition because different wood species are pulped. Tall oil is a mixture of fatty acids with about 2% of resin acid remaining after distillation. It is mainly oleic and linoleic acid (together 75–80%) and is the cheapest source of such acids used for industrial purposes. It also contains *pinolenic* acid and some conjugated diene acid. It is also a source of *phytosterols* used in the production of *Benecol™*.
A. Hase et al., *Tall oil as a fatty acid source,, Lipid Technology, 1994, 6, 110-114.*

tallow
animal edible tallow, normally obtained from beef but also from sheep and goats. Annual production is about 7 million tonnes. Tallow contains mainly saturated (60%, 16:0 and 18:0) and monounsaturated acids (40%, mainly 18:1 with some 16:1). Also present are odd-chain, *branched-chain*, and *trans fatty acids*, and *cholesterol* (0.08–0.14% in beef tallow and 0.23–0.31% in mutton tallow). Tallow is extensively used as a starting material for the production of *surfactants* and other oleochemicals.
J.A. Love, *Bailey's Industrial Oil and Fat Products (ed. Y.H. Hui) John Wiley & Sons, New York, 1996, Volume 1, pp.1-18.*

tariric acid
see *acetylenic acids*.

taurolipids
a range of taurine (2-aminoethanesulfonic acid) containing lipids which have been identified in bacteria.
K. Kaya, *Chemistry and biochemistry of taurolipids, Progress in Lipid Research, 1992, 31, 87-108*

taxoleic acid
trivial name of *5c9c-octadecadienoic acid*.

Tay-Sachs disease
characterised by an accumulation of a *ganglioside* (GM_2) in the brain due to defective enzymes (hexosaminidases).

Tay-Sachs ganglioside
a monosialoganglioside (GM$_2$). See *Tay-Sachs disease, gangliosides.*

TBA test
see *thiobarbituric acid.*

TBARS
thiobarbituric acid reactive substances. Compounds such as malonic dialdehyde and αβ-unsaturated aldehydes which react with *thiobarbituric acid.*
E.N. Frankel, Lipid Oxidation, The Oily Press, Dundee (1998)

TBHQ
see *tertbutyl hydroquinone.*

TCL
triple chain length, see *crystal structure.*

terpenoid lipid
isoprenoids such as *sterols, triterpenoids* and some *vitamins* which are biosynthetically derived from isoprene units through the mevalonate pathway.

tertiarybutylhydroquinone
a powerful synthetic antioxidant. Though widely used it is not a permitted food ingredient in Europe.

Tertiarybutylhydroquinone
2-*tert*-butyl-4-hydroxyphenol
C$_{10}$H$_{14}$O$_2$
Mol. Wt.: 166.22

tetracosanoic acid
the saturated C_{24} acid (lignoceric, mp 84.2°C) occurs along with its C_{20} and C_{22} homologues in *groundnut oil* (total 5–8%).

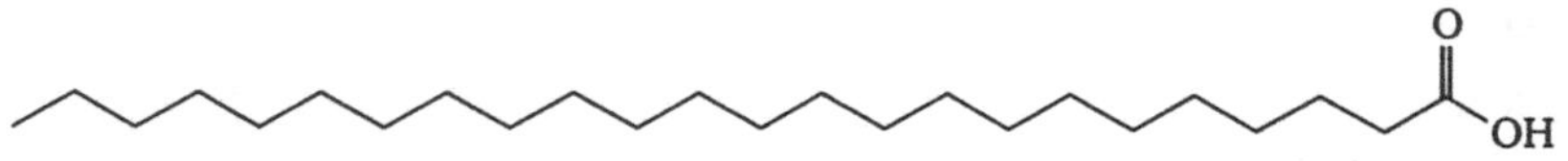

Lignoceric acid
Tetracosanoic acid
(24:0)
$C_{24}H_{48}O_2$
Mol. Wt.: 368.6

tetracosenoic acid
the *cis*-15 acid (nervonic, selacholeic) which melts at 41°C is an important component of cerebrosides and other sphingolipids. It is also present in many fish oils at low levels and in some vegetable fats such as honesty (*Lunaria*) and nasturtium (*Tropaeolum*) seed oils.

tetradecanoic acid
the C_{14} acid (*myristic*, mp 54.4°C, b.p. 149°C/2mm). It occurs in *lauric oils*, in many *animal fats* including *milk fats* and in most *fish oils*. It is considered to be the saturated acid with the greatest effect on increasing cholesterol levels in blood plasma.

tetradecenoic acid
cis-9-tetradecenoic acid (myristoleic acid 14:1) is present in the seed oil of *Lophira alata*. *Cis*-4-tetradecenoic acid (mp 18–18.5°C) occurs in *Lindera obtusiloba* and some other rare seed oils. It has a number of trivial names including obtusilic, tsuzuic and tsuduic.

TFA
see *trans fatty acids*.

TG
triglycerides. See *triacylglycerols*.

thapsic acid
trivial name for the C_{16} dibasic acid (hexadecanedioic acid, mp 126°C), which

occurs naturally in some waxes.

Thapsic acid
Hexadecanedioic acid
$C_{16}H_{30}O_4$
Mol. Wt.: 286.4

Theobroma cacao (Sterculia)
see *cocoa butter*.

theobroma oil
pharmacopoeial term for *cocoa butter*.

thermotropic mesomorphism
see *liquid crystals*.

thin-layer chromatography
a planar chromatographic separation technique using a thin layer of stationary phase, typically silica, spread on a plate of glass or metal. The plate is spotted with a solution at one end and is then developed in a chamber with a solvent system, the mobile phase, moving up the plate by capillary forces within the layer. Detection is normally done by charring or by spraying with fluorescent or destructive or specific reagent.

J.C. Touchstone, Review: Thin-layer chromatographic procedures for lipid separation, Journal of Chromatography B, 671, 1995, 169-195. W.W. Christie and G. Dobson, Thin-layer chromatography re-visited, Lipid Technology, 1999, 11, 64-66.

thiobarbituric acid test
used in a colorimetric test for lipid oxidation. The test is believed to measure malondialdehyde (a secondary oxidation product of polyenoic acids) but reaction also occurs with other aldehydes. The procedure has been criticised for its empirical nature but is, nevertheless, widely used. See *TBARS*.

E.N. Frankel, Lipid Oxidation, The Oily Press, Dundee (1998).

Thiobarbituric acid
$C_4H_4N_2O_2S$
Mol. Wt.: 144.2

thromboxanes
a family of eicosanoid compounds. Structurally they are heterocyclic (pyran) acids synthesised in mammals from arachidonic acid (*eicosatetraenoic acid*). Thromboxane A_2 and thromboxane B_2 are C_{20} metabolites of *PGH$_2$*.
N.M. Wilson, Lipid Synthesis and Manufacture, Sheffield Academic Press, Sheffield, 1999, pp.127-161.

Thromboxane B$_2$
9α,11,15S-Trihydroxythromboxa-5Z,13E-dien-1-oic acid
$C_{20}H_{34}O_6$
Mol. Wt.: 370.5

thynnic acid
hexacosahexaenoic acid, 26:6 (structure not certain but presumably the *n*–3 isomer). Present in some fish oils at low levels.

TLC
see thin-layer chromatography.

timnodonic acid
trivial name for *eicosapentaenoic acid*.

Tirtiaux process
commercial fractionation of oils using detergents to assist the separation of the *stearin* and *olein* fractions. See *Lanza fractionation*.
R.E. Timms, Lipid Technologies and Applications, Marcel Dekker, New York, 1997, pp.199-222.

tobacco seed oil
by-product from the harvesting of tobacco plants (*Nicotiana tabacum* and *Nicotiana rustica*, Solanaceae). The oil, which is rich in linoleic acid (ca 70%), may be used as an edible oil or for technical purposes (e.g. alkyd resins).

tocols

most plants contain a mixture of *tocopherols* and *tocotrienols* which show both vitamin (E) and antioxidant activity.

tocopherols

a series of substituted benzopyranols (methyl tocols) occurring in vegetable oils. The four major members are α (5,7,8-trimethyl), β (5,8-dimethyl), γ (7,8-dimethyl) and δ (8-methyl). The C_{16} side-chain linked to the pyran ring is saturated in the tocopherols but contains three double bonds in the tocotrienols.

During refining much of the natural antioxidant may be lost. The composition of the natural mixtures of tocols varies from source to source. α-Tocopherol shows the highest vitamin activity whilst the δ and γ tocopherols are the most active antioxidants. Natural tocopherols are generally used at levels up to 500 ppm along with *ascorbyl palmitate* (200–500 ppm) which enhances antioxidant activity through regeneration of spent tocopherol. Some vegetable oils already contain tocols at levels of 200–800 ppm so that added material has only a limited effect.

Three types of tocopherol are produced for sale: (i) natural tocopherols isolated from soybean, sunflower or other vegetable oils which are mixtures of the α, β, γ and δ compounds, (ii) natural mixtures which have been methylated to convert the β, γ and δ compounds to α tocopherol (the trimethyl compound) with higher vitamin E activity, and (iii) synthetic α-tocopherol made from trimethylhydroquinone and phytyl bromide. This is entirely the trimethyl compound but a mixture of eight stereoisomers. All three show antioxidant and vitamin E activity but only the last two can be sold as vitamin E.

Compound	R^1	R^2	R^3
α-Tocopherol	$-CH_3$	$-CH_3$	$-CH_3$
β-Tocopherol	$-CH_3$	-H	$-CH_3$
γ-Tocopherol	-H	$-CH_3$	$-CH_3$
δ-Tocopherol	-H	-H	$-CH_3$

J.P. Clark, Tocopherols and sterols from soybeans, Lipid Technology, 1996, 8, 111-114. K. Sundram and Abdul Gapor Md. Top, Vitamin E from palm oil; its extraction and nutritional properties, Lipid Technology, 1992, 4, 137-141. E.N. Frankel, Lipid Oxidation, The Oily Press, Dundee (1998). T. Netscher, Lipid Synthesis and Manufacture (ed. F.D. Gunstone) Sheffield Academic Press, Sheffield (1999) pp.250-267.

tocotrienols
A group of tocols related to tocopherol but having three unsaturated centres in the C_{16} side chain.

top olein
a commercial product obtained by repeated *fractionation* of *palm oil*. It has a cloud point <0°C and an iodine value of 70–72. The content of saturated acids is only 31% compared with values of 50 and 45 for palm oil and *palm olein* respectively.

E. Deffense, Dry multiple fractionation; trends in products and applications, Lipid Technology, 1995, 7, 34-38.

totox value
an empirical assessment of oxidative deterioration based on the *peroxide value* (primary oxidation products) and *anisidine value* (secondary oxidation products).

totox value = 2 peroxide value + anisidine value.

E.N. Frankel, Lipid Oxidation, The Oily Press, Dundee (1998).

tournesol (huile de)
French name for *sunflower* seed oil.

***trans* acids**
fatty acids with *trans (E)* unsaturated centres, in contrast to the *cis (Z)* configuration in most natural acids. Dietary *trans* acids come mainly from three sources: as a result of partial *hydrogenation* of vegetable or fish oils, in dairy fats where they are formed by *biohydrogenation* in the rumen, and through exposure to high temperatures during *refining*. They are present in a small number of seed oils. There has been concern regarding their adverse nutritional effects. They can also be formed from the cis isomers by *stereomutation*. They are higher melting than the *cis* isomers.

J.L. Sebedio and W.W. Christie, Trans Fatty acids in Human Nutrition, The Oily Press, Dundee (1998). D. Firestone and A. Sheppard, Advances in Lipid Methodology — One (ed. W.W. Christie) The Oily Press, Ayr (1992) pp.273-322. J.M. Vatele, Lipid Synthesis and Manufacture (ed. F.D. Gunstone) Sheffield Academic Press, Sheffield (1999) pp.1-45.

trans determination
normally refers to the measurement of the concentration of non-conjugated *trans* double bonds in fats and oils. This may be measured by *infrared spectroscopy*, *gas chromatography*, *silver ion chromatography*, or ^{13}C *nuclear magnetic resonance spectroscopy*. Standard methods are described by AOCS [Cd 14 61(89) and Cd 17 85(89)].

D. Firestone and A. Sheppard, Advances in Lipid Methodology — One (ed. W.W. Christie) The Oily Press, Ayr (1992) pp.273-322. R.E. McDonald and M.M. Mossoba, Food Lipids: Chemistry, Nutrition, and Biotechnology, (ed. C.C. Akoh and D.B. Min) Marcel Dekker, New York (1998) pp.137-166. J.L. Sebedio and W.W. Christie, Trans Fatty acids in Human Nutrition, The Oily Press, Dundee (1998).

trans double bond
see *unsaturation, trans acids.*

transesterification
method for producing esters from other esters under the influence of basic, acidic or enzymic catalysts. The esters may react with an alcohol (*alcoholysis, methanolysis, glycerolysis*), an acid (*acidolysis*), or another ester (*interesterification*).

F.D. Gunstone, Lipid Synthesis and Manufacture (ed. F.D. Gunstone) Sheffield Academic Press, Sheffield (1999) pp.321-346. A. Rosendaal and A.R. Macrae, Lipid Technologies and Applications (ed. F.D. Gunstone and F.B. Padley) Marcel Dekker, New York, (1997), pp.223-264.

transition temperature
normally refers to the temperature for (i) thermotropic transformations in polymorphic systems, e.g. fats (see *polymorphism*) and (ii) acyl melting points of lyotropic liquid crystalline systems, e.g. biological membranes (see *liquid crystals*).

transport flame-ionization detector
a detector for liquid chromatography based on the principle of trapping the column effluent on a wire or belt, which transports the eluted material, through an evaporating step to remove the solvent, to a combustion chamber and a flame-ionization detector. No detector of this type appears to have been produced commercially.

triacylglycerols
also called triglycerides, TAG, and TG. The triesters of glycerol, which may contain one, two or three different fatty acids, are the most common form of natural lipid. They occur widely in plants and animals in storage organs (depot

fats, adipocytes, etc) as mixtures of molecular species and are known commonly as fats and oils depending on whether they are solid or liquid at ambient temperature. They provide an efficient source of energy (9 kcal/g) and also serve for insulation and protective purposes. Individual compounds with specific acids in each position can be synthesised using appropriate protecting groups when this is necessary. The solid compounds show *polymorphism* because they exist in more than one crystalline form and hence have several melting points. The highest melting form of triacylglycerols with only one kind of fatty acid are tricaprin (33°C), trilaurin (46°C), trimyristin (56°C), tripalmitin (66°C), tristearin (73°C), triarachidin (78°C), tribehenin (83°C), triolein (5°C), trielaidin (41°C) and trierucin (32°C). Many compounds with two or three different acyl groups have also been synthesised.

Biosynthesis: *glycerol-3-phosphate* is converted to triacylglycerol through lysophosphatidic acid, *phosphatidic acid*, and 1,2-*diacylglycerol*. *Acylation* at each glycerol hydroxyl is effected by *acyl-coenzyme-A* or acyl-ACP in the presence of regiospecific enzymes.

Digestion/metabolism: under the influence of *lipases*, triacylglycerols are hydrolysed to free acids and 2-*monoacylglycerol*. These are reconverted to triacylglycerol for transport as *chylomicrons* or *lipoproteins*.

P.E. Sonnet, Lipid Synthesis and Manufacture (ed. F.D. Gunstone) Sheffield Academic Press, Sheffield (1999) pp.162-184.

trichosanic acid
punicic acid; trivial name for the conjugated octadecatrienoic acid, 9c11t13c-octadecatrienoic acid. See also *conjugated acids*.

tricosanoic acid
the C_{23} odd-chain saturated acid (mp 71°C). Its methyl ester is used as an internal standard in the gas chromatographic analysis of fish oils.

Tricosanoic acid
(23:0)
$C_{23}H_{46}O_2$
Mol. Wt.: 354.6

tridecanoic acid
the C_{13} odd-chain saturated acid (mp 45°C).

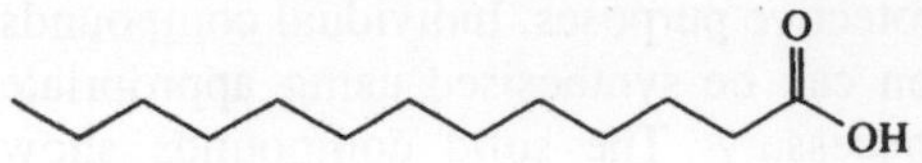

Tridecanoic acid
(13:0)
$C_{13}H_{26}O_2$
Mol. Wt.: 214.3

triglycerides
see *triacylglycerols.*

tripalmitin
glycerol 1,2,3-tripalmitate (mp 66°C). This is present in palm oil and can be synthesised from glycerol and palmitic acid.

trisialogangliosides
see *gangliosides.*

triterpenoids
see *terpenoid lipids.*

tritriacontanoic acid
ceromelissic acid, 33:0.

Trolox™
a derivative of α-tocopherol produced industrially in which the isoprenoid side-chain is replaced by a carboxyl group (2-carboxy-2,5,7,8-tetramethylchroman-6-ol), an antioxidant with more hydrophobic character than α-tocopherol.
E.N. Frankel, Lipid Oxidation, The Oily Press, Dundee (1998).

tsuduic acid
see *tetradecenic acid.*

tsuzuic acid
see *tetradecenic acid.*

tuberculostearic acid

10-methyloctadecanoic acid (*R*-form, mp 12.8–13.4°C). It is present in the lipids of the tubercle bacillus and in other bacterial lipids. See also *branched acids*.

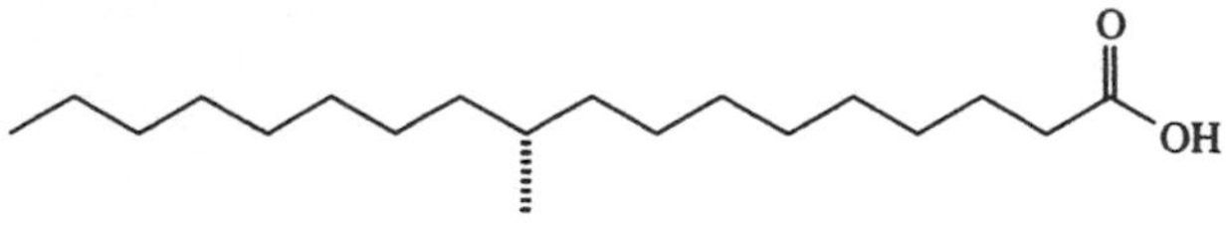

Tuberculostearic acid
10*R*-Methyloctadecanoic acid
$C_{19}H_{38}O_2$
Mol. Wt.: 298.5

tuna oil

a good source of *n*–3 polyene acids, particularly rich in *docosahexaenoic acid*. See also *fish oils*.

tung oil

also called China wood oil and produced from nuts of the trees *Aleurites fordii* and *A. montana*. It is a minor oil grown in China, Japan, and USA and characterised by its high content of *eleostearic acid*. It dries (hardens) even quicker than linseed oil and is used in enamels and varnishes.

Turkey-red oil

an early surface-active compound much used in textile spinning. Made industrially by sulphation of castor oil whereby the hydroxyl group is converted to a sulphate.

U

ultraviolet and visible spectra

spectra covering the range 200–750 nm. Most used in lipid studies to detect conjugated unsaturation: dienes at 230–240nm, trienes display a triple peak at 261, 271 and 281 nm. They are also used to detect hydroperoxidation during which conjugated unsaturated compounds are formed. End absorption occurs in the range 200–206 nm and this is exploited in the ultraviolet detector for HPLC systems.

undecenoic acid

10-undecenoic acid (mp 24.5°C) is produced along with *heptanal* by pyrolysis of *ricinoleic acid*. Its salts are used as antifungal agents and its esters as perfumery ingredients. It can be converted to 11-amino-undecenoic acid which is the monomer used to make nylon-11 (*Rilsan™*).

M.J. Caupin, Lipid Technologies and Applications, (ed. F.D. Gunstone and F.B. Padley) Marcel Dekker, New York, 1997, pp.787-795. L. Rebrovic and F.D. Gunstone, Oxidative cleavage of unsaturated fatty acids, Lipid Technology, 1996, 8, 135-137.

10-Undecenoic acid
$C_{11}H_{20}O_2$
Mol. Wt.: 184.3

unsaponifiable matter

material from a lipid sample which can be extracted by petroleum ether or diethyl ether after alkaline hydrolysis (e.g. hydrocarbons, alcohols, sterols etc). Excludes acids present as soaps and glycerol and other water-soluble compounds.

unsaturated acids

acids having one or more unsaturated centres. See *unsaturation, monoenoic acids, polyenoic acids, acetylenic acids.*

unsaturation

generally used to describe 4-electron (double bond) and 6-electron (triple bond) linkages between carbon atoms. See *olefinic, monoenoic acids, polyenoic acids, acetylenic acids.*

urea fractionation
the separation of fatty acids or alkyl esters by urea in methanol or methanol/ethanol solution. Straight-chain saturated compounds readily form urea adducts, which precipitate; unsaturated and branched-chain compounds do so less readily. It is used preparatively for the enrichment of polyenoic or branched-chain acids or esters from natural sources. These generally remain in the mother liquor. This process can be carried out on a gram to tonne scale.

urofuranic acids
furanoid acids of low molecular weight present in urine such as the one formulated. Probably metabolites of the *furanoid acids*.

R = $CH_3(CH_2)_4$-, $CH_3(CH_2)_2$-, $CH_3CH_2CH(OH)$- or CH_3CH_2CO-
Urofuranic acids

uropygial glands
sebaceous glands of birds. They produce branched-chain fatty acids incorporated into wax esters. These are spread on the feathers and serve as a barrier to water.

ustilagic acid
see *ustilic acid.*

ustilic acid
15,16-dihydroxyhexadecanoic acid, and 2,15,16-trihydroxy-hexadecanoic acid, components of the antibiotic ustilagic acid.

V

vaccenic acid
trivial name for 11-*octadecenoic acid* (usually the *trans* isomer).

valeric acid
the trivial name for pentanoic acid, 5:0.

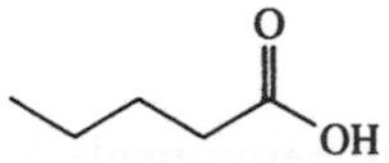

Valeric acid
pentanoic acid
(5:0)
$C_5H_{10}O_2$
Mol. Wt.: 102.1

vanaspati
an alternative to *ghee* that is based on vegetable oil. It is a grainy product made from *hydrogenated* oil (*groundnut, soybean, cottonseed, rice bran, rapeseed, or palm*) and widely used in India and Pakistan.
K.T. Achaya, Lipid Technologies and Applications (ed. F.D. Gunstone and F.B. Padley) Marcel Dekker, New York, (1997), pp.369-390.

vegetable oils
fatty oils of vegetable origin. Most (e.g. soya, rape, sunflower) are expressed from the seeds but others occur in the soft fleshy fruit (endosperm) such as palm, olive and avocado. They contain phytosterols such as brassicasterol, campesterol, stigmasterol, β-sitosterol, and Δ5- and Δ7-avenasterols. The major sterols (campesterol, stigmasterol and sitosterol) are present at levels of 0.5–1.1% in rapeseed oil, 0.2–0.4% in soybean oil, and 0.3–0.5% in sunflower oil.

vernolic acid
the best known of the natural epoxy acids. It is *cis*-12,13-epoxy-*cis*-9-octadecenoic acid and occurs in vernonia oils such as that from *Vernonia galamensis* (~75%). Attempts are being made to develop this as a commercial crop and also *Euphorbia lagascae* (60-65% vernolic acid). Such oils have a

number of potential uses. See also *leukotoxins*.

Vernolic acid
12*S*,13*R*-Epoxy-9*Z*-octadecenoic acid
$C_{18}H_{32}O_3$
Mol. Wt.: 296.4

Vernonia galamensis
its seed oil is rich in *vernolic acid* (ca 75%). Attempts are being made to develop this as a commercial crop. See also *epoxy acids* and *Euphorbia lagascae*.

very-low-density lipoproteins
VLDL. See *lipoproteins*.

VHEAR
very-high-erucic rapeseed oil with at least 66% of *erucic acid*.

vinyl ethers
see *plasmalogens*.

virgin olive oil
see *olive oil (grades)*.

vitamins
dietary substances required from the diet in small amount for the normal functioning of the body. Important fat-soluble vitamins are vitamin A (*retinol*), vitamin E (*tocopherol*), vitamin D (*cholecalciferol*) and vitamin K (*menaquinone*). Refining processes may lead to a reduction in the level of these materials in vegetable oils but sometimes they can be recovered from minor fractions. Partial hydrogenation can also affect these compounds.

VLDL
see *lipoproteins*.

volicitin
amide of 17-hydroxylinolenic acid and L-glutamine. It is a chemical elicitor,

present in insect saliva, which triggers plants to release a mixture of volatile compounds. These attract parasites and predatory insects to the herbivore.

von Rudloff oxidation
oxidation of olefinic compounds by potassium permanganate and sodium periodate resulting in fission and giving short-chain acids. The procedure was used to determine double bond positions in natural fatty acids. See also *ozonolysis*.

W

walnut oil
oil from walnuts *(Juglans regia)*. It is a linoleic-rich oil (ca 60%) and also contains α-linolenic acid (8%). Walnut oil is rich in *tocopherols* and shows high vitamin E activity.

waxes
water-resistant materials made up of hydrocarbons, long-chain acids and alcohols, *wax esters* and other long-chain compounds. They are produced by animals *(beeswax, wool wax* (lanolin), *sperm whale wax* and *orange roughy oil)* and plants (candelilla, *carnauba,* rice-bran, sugar cane and *jojoba)* and may be solid or liquid. All leaf surfaces are covered by a microcrystalline layer of wax. They are used in the food, pharmaceutical and cosmetic industries and for the protection of surfaces (automobiles, furniture, paper etc.). Attempts are being made to develop a rape plant which produces wax esters in its seeds lipids.

S. Puleo and T.P. Rit. Natural waxes; past, present, and future, Lipid Technology, 1992, 4, 82-90. S. Li and E.J. Parish, Food Lipids: Chemistry, Nutrition, and Biotechnology, (ed. C.C. Akoh and D.B. Min) Marcel Dekker, New York (1998) pp.89-114. G. Bianchi, Waxes: Chemistry, Molecular Biology and Functions: Plant Waxes, (ed. R. J. Hamilton) Oily Press, Dundee, (1995) 175-222.

wax esters
esters of long-chain acids and long-chain alcohols which are important components of *waxes.*

whale oils
see *spermaceti.*

wheat germ oil
oil from wheat germ (the embryo of the seed of *Triticum aestivum).* It is rich in linoleic acid (ca 60%) and also contains α-linolenic acid (ca 5%). The oil is rich in tocopherols and shows high vitamin E activity.

Wijs' reagent
a solution of iodine monochloride in acetic acid (usually 0.2 mol/litre) used to measure the *iodine value.*

winterisation
a commercial process for the separation of oils and fats into solid *(stearin)* and

liquid (*olein*) fractions by crystallization at sub-ambient temperatures. It was applied originally to cottonseed oil to produce a solid-free product.

woolwax

is recovered from wool during the scouring process. It contains *wax esters* (48–49%), *sterol esters* (32–33%), *lactones* (6–7%), triterpene alcohols (4–6%), free acids (3–4%) and free *sterols* (1%). The fatty acids in the wax esters are straight-chain and branched-chain (*iso* and *anteiso*) with some hydroxy acids. It is also known as lanolin.

S. Orr, A brief guide to lanolin technology and applications, Lipid Technology, 1998, 10, 10-14.

R.J. Hamilton, Waxes: Chemistry, Molecular Biology and Functions, (ed. R. J. Hamilton), The Oily Press, Dundee, (1995) p.257.

X

xenobiotic lipids
fat-soluble compounds in the environment which enter into the food chain and may become associated with lipid depots in tissues by reason of their solubility or by chemical bonding.

J.L. Harwood, The Lipid Handbook, 2nd edition (ed. F.D. Gunstone, J.L. Harwood, and F.B. Padley), Chapman and Hall, London, 1999, pp.700-705. P.F. Dodds, Xenobiotic lipids: the inclusion of xenobiotic compounds in pathways of lipid biosynthesis, Progress in Lipid Research, 1995, 34, 219-247.

ximenic acid
17-*cis*-hexacosenoic acid, 17c-26:1. Present in the seed oil from *Ximenia americana.*

ximenynic acid
trivial name for a solid acid (mp 39–40°C) with conjugated enyne unsaturation 9a11t-18:2. Occurs in high levels in *Santalum acuminatum* seed oil and in the seed lipids of many species of Santalaceae and Oleaceae. Also called santalbic acid.

X-ray diffraction
study of the diffraction of a beam of X-rays by solid crystals or thin sections. The technique is used in the investigation of the arrangement of lipid molecules, multiple melting phenomena and identification of lipid aggregates, e.g. lamellar or hexagonal structures.

P. Laggner, Spectral Properties of Lipids (eds. R.J. Hamilton and J. Cast), Sheffield Academic Press, Sheffield (1999), pp.327-367.

Y

yellow fats
the general name for yellow spreading fats including *butter, margarine* and low-fat spreads.

Yukagaku
journal of the Japan Oil Chemists' Society.

Z

Zea mays (Graminae, grass family)
see corn oil.

Zenith process
a *refining* procedure involving *degumming, neutralization* with dilute alkali, and *bleaching* which has some advantages over the more conventional procedures.
D.A. Allen, Lipid Technologies and Applications (ed. F.D. Gunstone and F.B. Padley) Marcel Dekker, New York (1997), pp.137-167.

zoomaric acid
see 9-*hexadecenoic acid.*

Journals and Periodicals

Advances in Lipid Research
ACADEMIC PRESS, New York. From 1963.

Biochimica Biophysica Acta - Biomembranes
ELSEVIER, Amsterdam. From 1947.

Chemistry and Physics of Lipids
ELSEVIER, Shannon. From 1966.

Current Opinion in Lipidology
CURRENT SCIENCE, Philadelphia. From 1990.

European Journal of Lipid Science and Technology
WILEY-VCH, Weinheim. From 2000.

Fat Science Technology
(Fette Seifen Anstrichmittel), KONRADIN INDUSTRIVERLAG GmbH, Hamburg. From 1899 to 1995 (see Fett/Lipid).

Fett/Lipid
WILEY-VCH, Weinheim. From 1996 to 1999 (see *European Journal of Lipid Science and Technology*).

Grasas Y Aceites
INSTITUTO DE LA GRASA (CSIC) Sevilla, Spain.

INFORM
AOCS PRESS, Champaign, Illinois. From 1990.

Journal of Lipid Mediators
ELSEVIER, Amsterdam. From 1989.

Journal of Lipid Research
LIPID RESEARCH INC, New York. From 1959.

Journal of Liposome Research
MARCEL DEKKER, New York. From 1989.

Journal of Surfactants and Detergents
AOCS PRESS, Champaign, Illinois. From 1998.

Journal of the American Oil Chemists' Society
AOCS PRESS, Champaign, Illinois. From 1923.

Journal of the Oil Technologist's Association of India
OIL TECHNOLOGIST'S ASSOC'N OF INDIA, Bombay. From 1969.

La Rivista Italiana delle Sostanze Grasse
STAZIONE SPERIMENTALE DEGLI OLI E DEI GRASSI, Milano.
From 1923.

Lipids
AOCS PRESS, Champaign, Illinois. From 1966.

Lipid Technology
PJ BARNES & ASSOCIATES, Bridgwater. From 1989.

Lipid Technology Newsletter
PJ BARNES & ASSOCIATES, Bridgwater. From 1995.

Oils and Fats International
DMG BUSINESS PUBLICATIONS, Redhill. From 1988.

Oilseeds
SATSOUTH LTD, London

Oil World
ISTA Mielke GmbH, Hamburg. From 1958.

Oleagineux
INSTITUT DE RECHERCHES HUILES ET OLEAGINEUX, Paris.
From 1946 to 1993 (see *Oleagineux Corps gras Lipids*).

Oleagineux Corps Gras Lipides
INSTITUT DE RECHERCHES HUILES ET OLEAGINEUX, Paris.
From 1994.

Progress in Lipid Research
PERGAMON, Ne;w York. From 1952.

Raps
VERLAG TH. MANN, Gelsenkirchen Nordring Buer. From 1983.

Revue Française des Corps Gras
Paris. From 1954 to 1993 (see *Oleagineux Corps gras Lipides*)

Seifen, Öle, Fette, Wachse
PRESSE-DRUCK-UND-VERLAGS GmbH, Augsburg. From 1975.

Yukagaku
JAPAN OIL CHEMIST'S SOCIETY, Tokyo. From 1951.

Book List

1. **The Chemical Constitution of Natural Fats.** 4th Edn. Hilditch T.P., Williams P.N., CHAPMAN & HALL, 1964.
2. **Methods in Enzymology. XIV. Lipids.** Lowenstein J.M., Ed., ACADEMIC PRESS, 1965.
3. **Plant Lipid Biochemistry.** Hitchcock Ç., Nichols B.W., ACADEMIC PRESS, 1971.
4. **Analysis of Triglycerides.** Litchfield C., ACADEMIC PRESS, 1972.
5. **Methods in Enzymology. XXXV. Lipids.** Lowenstein J.M., Ed., ACADEMIC PRESS, 1975.
6. **Lipid Chromatographic Analysis.** Vol. 1 and 2. Second Edn. Marinetti G.V., Ed., MARCEL DEKKER, 1976.
7. **Chemistry and Biochemistry of Natural Waxes.** Kolattukudy P.E., Ed., ELSEVIER, 1976.
8. **Cholesterol.** Sabine J.R., MARCEL DEKKER, 1977.
9. **Handbook of Lipid Research.** Vol. 1. Fatty Acids and Glycerides. Kuksis A., Ed., PLENUM PRESS, 1978.
10. **The Pharmacological Effect of Lipids,** Kabara J.J., Ed., AOCS PRESS, 1978.
11. **Fatty Acids.** Pryde E.H., Ed., AOCS PRESS 1979.
12. **Geometrical and Positional Fatty Acid Isomers** Emken E.A., Dutton H.J., Eds, AOCS PRESS 1979.
13. **Fats and Oils: Chemistry and Technology.** Hamilton R.J., Bhati, A. Eds, APPLIED SCIENCE PUBLISHERS, 1980.
14. **Lipid Biochemical Preparations.** Bergelson L.D., Ed., ELSEVIER, 1980.
15. **Lipids in Evolution.** Nes W.R., New W.D., PLENUM PRESS, 1980.
16. **Vitamin E: A Comprehensive Treatise,** Machlin L.J., Ed., MARCEL DEKKER, 1980.
17. **Membrane Structure.** Finean J.B., Michell R.H., Eds, ELSEVIER, 1981.
18. **The Biology of Cholesterol and Related Steroids.** Myant N.B., HEINEMANN, 1981.
19. **Lipid Metabolism in Ruminant Animals.** Christie W.W., Ed., PERGAMON PRESS, 1981.
20. **New Sources of Fats and Oils.** Pryde E.H., Princen L.H., Mukherjee K.D., Eds, AOCS PRESS 1981.
21. **Fatty Alcohols, Raw Materials, Methods, Uses.** Henkel. Ed., HENKEL, Düsseldorf, 1982.
22. **Lipid Analysis.** 2nd Edn. Christie W.W., PERGAMON PRESS, 1982.
23. **Phospholipids.** Hawthorne J.N., Ansell G.B., Eds, ELSEVIER, 1982.
24. **Phospholipids in the Nervous System.** Volume 1. **Metabolism.** Horrocks L.A., Ansell G.B., Porcellatio G., Eds, RAVEN PRESS, 1982.

25. **Liposome Letters.** Bangham A.D., Ed., ACADEMIC PRESS, 1983.

26. **New Vistas in Glycolipid Research.** Makita A., Taketomi T., Handa S., Nagai Y., Eds, PLENUM PRESS, 1983.

27. **The Lively Membranes.** Robertson R.N., CAMBRIDGE UNIVERSITY PRESS, 1983.

28. **Ether Lipids, Biochemical and Biomedical Aspects,** Mangold H.K., Paltauf F., Eds, ACADEMIC PRESS, 1983.

29. **Developments in Dairy Chemistry. 2. Lipids.** Fox P.F., Ed., ELSEVIER, 1983.

30. **Handbook in Lipid Biochemistry. Volume 3. Sphingolipid Biochemistry.** Kanfer J.N., Hakomori S.I., Eds, PLENUM PRESS, 1983.

31. **Lipids in Cereal Technology.** Barnes P.J., Ed., ACADEMIC PRESS, 1983.

32. **A Guidebook to Lipoprotein Technique.** Mills G.L., Lane P.A., Weech P.K., ELSEVIER, 1984.

33. **Fats in Animal Nutrition.** Wiseman J., Ed., BUTTERWORTHS, 1984.

34. **Handbook of Chromatography. Lipids.** Volumes 1 and 2. Mangold H.K., Ed., CRC PRESS, 1984.

35. **Biological Membranes.** Volume 4. Chapman D., Ed., ACADEMIC PRESS, 1984.

36. **Biochemistry of Lipids and Membranes.** Vance D.E., Vance J.E., Eds, THE BENJAMIN/CUMMINGS PUBLISHING COMPANY, 1985.

37. **Fat-soluble Vitamins – Their Biochemistry and Applications.** Diplock A.T., Ed., HEINEMANN. 1985.

38. **Flavor Chemistry of Fats and Oils.** Min D.B., Smouse T.H., Eds, AOCS PRESS, 1985.

39. **Glycolipids.** New Comprehensive Biochemistry. Volume 10, Wiegandt H., Ed., ELSEVIER, 1985.

40. **Handbook of Soy Oil Processing and Utilisation.** Erickson D.R., Pryde E.H., Brekke O.L., Mounts T.L., Falb R.A., Eds, AOCS PRESS, 1985.

41. **Jojoba: New Crop for Arid Lands, New materials for Industry.** National Research Council, NATIONAL ACADEMY PRESS, 1985.

42. **Lecithins,** Szuhaj B.F., List G.R., Eds, AOCS PRESS, 1985.

43. **Methods for Nutritional Assessment of Fats,** Beare-Rogers J., Ed., AOCS PRESS, 1985.

44. **Phospholipids in Nervous Tissues,** Eichberg J., Ed., WILEY-VCH, 1985.

45. **Sterols and Bile Acids,** New Comprehensive Biochemistry. Volume 12, Danielsson H., Sjövall J., Eds, ELSEVIER, 1985.

46. **The Adaptive Role of Lipids in Biological Systems,** Hadley N.F., JOHN WILEY & SONS, 1985.

47. **The Pharmacological Effect of Lipids II.,** Kabara J.J., Ed., AOCS PRESS, 1985.

48. **Analysis of Oils and Fats,** Hamilton R.J., Rossell J.B., Eds, ELSEVIER, 1986.

49. **Fat Absorption.** Volumes 1 and 2., Kuksis E., Ed., CRC PRESS, 1986.

50. **Fish and Human Health,** Lands W.E.M., ACADEMIC PRESS, 1986.

51. **Lipids and Membranes. Past, Present and Future,** Op Den Kamp J.A.F., Roelofsen B., Wirtz K.W.A., Eds, ELSEVIER, 1986.

52. **Lipids, Chemistry, Biochemistry and Nutrition,** Mead J.F., Alfin-Slater R.B., Howton D.R., Popjak G., PLENUM PRESS, 1986.

53. **Methods in Enzymology. Volume 128. Plasma Lipoproteins. Preparations, Structure and Molecular Biology,** Albers J.J., Segest J.P., Eds, ACADEMIC PRESS, 1986.

54. **Methods in Enzymology. Volume 129. Plasma Lipoproteins. Characterization, Cell Biology and Metabolism,** Albers J.J., Segest J.P., Ed., ACADEMIC PRESS, 1986.

55. **Phosphoinositides and Receptor Mechanisms,** Receptor Biochemistry and Methodology. Vol. 7, Putney J.W., Ed., ALAN R. LISS, 1986.

56. **Surfactants in Consumer Products: Theory, Technology and Application,** Falbe J., Ed., SPRINGER VERLAG, 1986.

57. **Techniques of Lipidology, Isolation, Analysis,** Kates M., ELSEVIER, 1986.

58. **The Physical Chemistry of Lipids. From Alkanes to Phospholipids. Handbook of Lipid Research 4,** Small D.M., PLENUM PRESS, 1986.

59. **Lecithin Der Unvergleichliche Wirkstoff,** Schäfer W., Wywiol V., ALFRED STROTHE VERLAG, 1986 (in German).

60. **Autoxidation of Unsaturated Lipids,** Chan H.W.S., Ed., ACADEMIC PRESS, 1987.

61. **Biological Membranes – Practical Approach,** Findlay J.B.C., Evans W.H., IRL PRESS, 1987.

62. **Biomembrane and Receptor Mechanisms,** Bertoli E., Chapman D., Cambria A., Scapagnini U., Eds, SPRINGER VERLAG, 1987.

63. **Chromatography of Lipids in Biomedical Research and Clinical Diagnosis,** Journal of Chromatography Library, Volume 37, Kuksis A., Ed., ELSEVIER, 1987.

64. **Handbook of Lipid Research. Volume 5. The Phospholipases,** Waite M., PLENUM PRESS, 1987.

65. **HPLC and Lipids. A Practical Guide,** Christie W.W., PERGAMON PRESS, 1987.

66. **Inositol Lipids in Cellular Signaling Proceedings,** Michell R.H., Putney J.W.Jr., Eds, COLD SPRING HARBOUR LABORATORY 1987.

67. **Lecithin Technical, Biological and Therapeutic Aspects,** Hanin I., Ansell G.B., Eds, PLENUM PRESS, 1987.

68. **Lipoprotein Lipase,** Borensztajn J., Ed., VENER PUBLISHER 1987.

69. **Liposomes From Biophysics to Therapeutics,** Ostro M.J., Ed., MARCEL DEKKER, 1987.

70. **Palm Oil,** Gunstone F.D., Ed., JOHN WILEY & SONS, 1987.

71. **Plasma Lipoproteins,** Gotto A.M., Ed., ELSEVIER, 1987.

72. **Platelet-Activating Factor and Related Lipid Mediators,** Snyder F., Ed., PLENUM PRESS, 1987.

73. **Seafoods and Fish Oils in Human Health and Disease,** Kinsella J.E., MARCEL DEKKER, 1987.

74. **The Biochemistry of Plants. Volume 9. Lipids: Structure and Function,** Stumpf P.K., Ed., ACADEMIC PRESS, 1987.

75. **The Chemistry and Technology of Jojoba Oil,** Wisniak J., AOCS PRESS, 1987.

76. **The Membranes of Cells,** Yeagle P., ACADEMIC PRESS, 1987.

77. **Baillère's Clinical Gastroenterology Nutritional Support. International Practice and Research,** Burns H.J.G., Ed., BAILLÈRE TINDALL, 1988.

78. **Biologically Active Ether Lipids,** Braque P., Mangold H.K., Vargaftig B.B., Eds, KRAGER, 1988.

79. **Biotechnological Application of Lipid Microstructures,** Gaber B.P., Schnur J.M., Chapman D., Eds, PLENUM PRESS, 1988.

80. **Crystallization and Polymorphism of Fats and Fatty Acids,** Garti N., Sato K., Eds, MARCEL DEKKER, 1988.

81. **Fat-Soluble Vitamin Assays in Food Analysis. A Comprehensive Review,** Ball G.F.M., ELSEVIER, 1988.

82. **Food – The Chemistry of Its Components.** 2nd Edn. Coultate T.P., ROYAL SOCIETY OF CHEMISTRY, 1988.

83. **Ganglioside,** Krämer G., Hopf H.C., Sandhoff K., Wiegandt H., Eds, GEORGE THIEME VERLAG, 1988.

84. **Langmuir-Blodgett Films 3.** Volume 1 and 2, Möbious D., Ed., ELSEVIER, 1988.

85. **Lipid Peroxidation in Biological Systems,** Proceedings, Sevanian E., Ed., AOCS PRESS PRESS, 1988.

86. **Lipids and Related Compounds.** Neuromethods 7, Boulton A.A., Baker G.B., Horrocks L.A., Eds, THE HUMANA PRESS, 1988.

87. **Microbial Lipids,** Volume 1, Ratledge C., Wilkinson S.G., Eds, ACADEMIC PRESS, 1988.

88. **Plant Membranes – Structure, Assembly and Function,** Harwood J.L., Walton T.J., Eds, PORTLAND PRESS, 1988.

89. **Seventh International Conference on Jojoba and its Uses,** Baldwin A.R., Ed., AOCS PRESS, 1988.

90. **Single Cell Oil,** Moreton R.S., Ed., LONGMAN SCIENTIFIC & TECHNICAL, 1988.

91. **The Basics of Industrial Oleochemistry,** Dieckelmann G., Heinz H.J., DIECKELMANN & HEINZ, 1988.

92. **The Biology of Surfactants,** Hills B.A., CAMBRIDGE UNIVERSITY PRESS, 1988.

93. **The Oil Palm,** Hartley C.W.S., LONGMAN, 1988.

94. **A Handbook of Hyperlipidemia,** Thompson G.R., CURRENT SCIENCE, 1989.

95. Analysis of Sterols and Other Biologically Significant Steroids, Nes W.D., Parish E.J., Eds, ACADEMIC PRESS, 1989.

96. Biomembranes. Molecular Structure and Function, Gennis R.B., SPRINGER VERLAG, 1989.

97. Brain Messengers and the Pituitary, Müller E.E., Nistico G., ACADEMIC PRESS, 1989.

98. Clinical Endocrinology and Metabolism. Lipoprotein Metabolism, Shepherd J., Ed., BAILLERE TINDALL, 1989.

99. Fatty Acids in Industry, Johnson R.W., Fritz E., Eds, MARCEL DEKKER, 1989.

100. Flavor Chemistry of Lipid Foods, Min D.B., Smouse T.H., Eds, AOCS PRESS, 1989.

101. Gas Chromatography and Lipids. A Practical Guide, Christie W.W., THE OILY PRESS, 1989.

102. Handling and Storage of Oilseed, Oils, Fats and Meal, Patterson H.B.W., ELSEVIER APPLIED SCIENCE, 1989.

103. Lecithins; Sources, manufacture & Uses, Szuhaj B.F., Ed., AOCS PRESS, 1989.

104. Lipid and Biopolymer Monolayers at Liquid Interfaces, Birdi K.S., PLENUM PRESS, 1989.

105. Lipids. What For? Where From? How Much?, Nestec, NESTEC, Lausanne, 1989.

106. Liposomes – a Practical Approach, New R.R.C., Ed., IRL PRESS, 1989.

107. Microbial Lipids, Volume 2, Ratledge C., Wilkinson S.G., Eds, ACADEMIC PRESS, 1989.

108. New Crops for Food and Industry, Wickens G., Ed., CHAPMAN & HALL, 1989.

109. Oil Crops of the World, Their Breeding and Utilisation, Röbbelen G., Downey R.K., Ashri A., Eds, McGRAW-HILL, 1989.

110. Surfactants in Solutions, Volume 7–10, Mittal K.L., Ed., PLENUM PRESS, 1989.

111. The Chemistry and Technology of Edible Oils and Fats, Hoffman G., ACADEMIC PRESS, 1989.

112. The Lipids of Human Milk, Jensen R.G., CRC PRESS, 1989.

113. The Role of Fats in Human Nutrition. 2nd Edn. Vergroesen A.J., Crawford M., ACADEMIC PRESS, 1989.

114. The Pharmacological Effects of Lipids III, Kabara J.J., Ed., AOCS PRESS, 1989.

115. Treatment of Hyperlipidemia, Strandberg K., Beermann B., Lönnerholm G., Eds, SOCIALSTYRELSEN 1989.

116. Applications of Lipases as Biocatalysts, Hammond R.C., Gotfredsen S.E., Eds, HARWOOD ACADEMIC PUBLISHERS, 1990.

117. **Bioactive Compounds from Plants,** Chadwick D.J. and Marsh J., Eds, JOHN WILEY & SONS, 1990.

118. **Cholesterol Metabolism, LDL, and the LDL Receptor,** Myant N.B., ACADEMIC PRESS, 1990.

119. **Cottonseed Oil,** Jones L.A., King C.C., Eds, NATIONAL COTTONSEED PRODUCTS ASSOCIATION, 1990.

120. **Disorders of Lipid Metabolism,** Marinetti G.V., PLENUM PRESS, 1990.

121. **Fish Oils in Nutrition,** Stansby M.E., Ed., CHAPMAN & HALL, 1990.

122. **Food Antioxidants,** Hudson B.J.F., Ed., ELSEVIER APPLIED SCIENCE, 1990.

123. **Food emulsions.** 2nd Edn. Revised and Expanded., Larsson K., Friberg S.E., Eds, MARCEL DEKKER, 1990.

124. **Handbook of Lipid Research 6. Glycolipids, Phosphoglycerides and Sulfoglycolipids,** Kates M., Ed., PLENUM PRESS, 1990.

125. **Lipids and Lipid Disorders,** Feher M.D., Richmond W., GOWER MEDICAL PUBLISHING, 1990.

126. **Lipids, Membranes and Aspects of Photobiology,** Methods in Plant Biochemistry. Vol.4., Harwood J.L., Bowyer J.R., Eds, ACADEMIC PRESS, 1990.

127. **Lipoprotein(a).** Scanu A.M., Ed., ACADEMIC PRESS, 1990.

128. **Marine Biogenetic Lipids, Fats and Oils.** Volumes 1 and 2. Ackman R.G., Ed., CRC PRESS, 1990.

129. **Methods in Inositide Research,** Irvine R.F., Ed., RAVEN PRESS, 1990.

130. **Olive Oil,** Kiritsakis A.K., Ed., AOCS PRESS, 1990.

131. **Plant Lipid Biochemistry, Structure and Utilization,** Quinn P., Harwood J.L., Ed., PORTLAND PRESS, 1990.

132. **Rapeseed, Chemistry and Technology,** Niewiadomski H., ELSEVIER, 1990.

133. **Recent Developments in the Technology of Surfactants,** Porter M.R., Ed., ELSEVIER APPLIED SCIENCE, 1990.

134. **Analysis of Oilseeds, Fats and Fatty Foods,** Rossel J.B., Pritchard J.L.R., Eds, ELSEVIER SCIENCE PUBLISHERS, 1991.

135. **Biotechnology of Plant Fats and Oils,** Rattray J.B.M., Ed., AOCS PRESS, 1991.

136. **Canola and Rapeseed,** Shahidi F., CHAPMAN & HALL, 1991.

137. **Glycerine. A Key Cosmetic Ingredient,** Jungermann E., Sonntag N.O.V., Eds, MARCEL DEKKER, 1991.

138. **Health Effects of Dietary Fatty Acids,** Nelson G.J., Ed., AOCS PRESS, 1991.

139. **Introduction to Fats and Oils Technology,** Wan P.J., Ed., AOCS PRESS, 1991.

140. **Lipid Biochemistry. An Introduction.** 4th Edn. Gurr M.I., Harwood J.L, CHAPMAN & HALL, 1991.

141. **Membrane Fusion,** Wilschut J., Hoekstra D., Ed., MARCEL DEKKER, 1991.

142. **Phospholipids: Biochemical, Pharmaceutical, and Analytical Considerations,** Hanin I., Pepeu G., Eds, PPÖ, 1991.

143. **Phospholipases.** Methods in Enzymology. Vol.197, Dennis E.A., Ed., ACADEMIC PRESS, 1991.

144. **The Handbook of Surfactants,** PORTER, M.R., BLACKIE, 1991.

145. **Understanding Your Cholesterol,** Yeagle P.L., ACADEMIC PRESS, 1991.

146. **Advances in Lipid Methodology – One,** Christie W.W., Ed., OILY PRESS, 1992.

147. **Determination of Vitamin E: Tocopherols and Tocotrienols,** Bourgeois C., ELSEVIER APPLIED SCIENCE, 1992.

148. **Role of Fats in Food and Nutrition.** 2nd Edn. Gurr M.I., ELSEVIER APPLIED SCIENCE, 1992.

149. **Advances in Lipid Methodology – Two,** Christie W.W., Ed., OILY PRESS, 1993.

150. **Chromatography for the Analysis of Lipids,** Hammond E.W., CRC PRESS 1993.

151. **Handbook of Derivatives for Chromatography,** 2nd Edn. Blau K., Halket J.M., Eds, JOHN WILEY & SONS, 1993.

152. **Designer Oil Crops: Breeding, Processing and Biotechnology,** Murphy D.J., Ed., VCN, 1993.

153. **Seed Oils for the Future,** MacKenzie S.L., Taylor D.C., Eds, AOCS PRESS, 1993.

154. **Diet and Heart Disease,** Ashwell M., Ed., BRITISH NUTRITION FOUNDATION, 1993.

155. **The Lipid Handbook,** 2nd Edn. Gunstone F.D., Harwood J.L., and Padley F.B., Eds, CHAPMAN & HALL, 1994.

156. **Developments in the Analysis of Lipids,** Tyman J.H.P., and Gordon M.H., Ed., ROYAL SOCIETY OF CHEMISTRY, 1994.

157. **New Trends in Lipid and Lipoprotein Analysis,** Sebedio J.L., and Perkins E.G., Eds, AOCS PRESS, 1994.

158. **Technological Advances in Improved and Alternative Sources of Lipids,** Kamel B.S., and Kakuda Y., BLACKIE ACADEMIC AND PROFESSIONAL 1994.

159. **Lipids: Molecular Organization, Physical Functions and Technical Applications,** Larsson Kåre, THE OILY PRESS 1994.

160. **Developments in Oils and Fats,** Hamilton R.J., Ed., CHAPMAN & HALL, 1995.

161. **Phospholipids; Characterization, Metabolism, and Novel Biological Applications,** Cevc G., Paltauf F., Eds, AOCS PRESS, 1995.

162. **Steroid Analysis,** Malkin H.L.J., Gower D.B., and Kirk B.N., Eds, BLACKIE 1995.

163. **Nouveau Dictionnaire des Huiles Vegetales,** Ucciani E., LAVOISIER TEC DOC, 1995.

164. **Waxes: Chemistry, Molecular Biology and Functions,** Hamilton R.J., Ed., THE OILY PRESS, 1995.

165. Brain Development: Relationship to Dietary Lipid and Lipid Metabolism, Jumpsen J., Clandinin M.T., Eds, AOCS PRESS, 1995.

166. Food Oils and Fats: Technology, Utilization, and Nutrition, Lawson H., CHAPMAN & HALL, 1995.

167. γ-Linolenic Acid: Metabolism and its Role in Nutrition and Medicine, Huang Y.-S., and Mills D.E., Eds, AOCS PRESS, 1995.

168. Nutrition, Lipids, Health and Disease, Ong A.S.H., Niki E., Packer L., Eds, AOCS PRESS, 1995.

169. Omega-3 fatty Acids and Health, Nettleton J.A., CHAPMAN & HALL, 1995.

170. Plant Lipid Metabolism, Kader J.-C., Mazliak P., Ed., KLUWER ACADEMIC, 1995.

171. Trans Fatty Acids, Anon BRITISH NUTRITION FOUNDATION, 1995.

172. Development and Processing of Vegetable Oils for Human Nutrition, Przybylski R., McDonald B.E., Eds, AOCS PRESS, 1995.

173. Handbook of Milk Composition, Jensen, R.G., Ed. ACADEMIC PRESS, 1995.

174. Bailey's Industrial Oil and Fat Products, 5 volumes, 5th Edn. Hui Y.H., JOHN WILEY & SONS, 1996.

175. Fatty Acid and Lipid Chemistry, Gunstone F.D., CHAPMAN & HALL, 1996.

176. Oils and Fats Manual, A Comprehensive Treatise, 2 volumes, Karleskind A., Ed., LAVOISIER TEC DOC, 1996.

177. Biochemistry of Lipids, Lipoproteins, and Membranes, Vance D .E., Vance J.E., Eds, ELSEVIER, 1996.

178. Advances in Lipid Methodology – Three, Christie W.W., Ed., OILY PRESS, 1996.

179. Olive Oil: Chemistry and Technology, Boskou E., Ed., AOCS PRESS, 1996.

180. Safflower, Smith J.R., AOCS PRESS, 1996.

181. Deep Frying: Chemistry, Nutrition and Practical Applications, Perkins E.G., and Erickson M.D., Eds, AOCS PRESS, 1996.

182. Food Lipids and Health, McDonald R.E., and Min D.B., Ed., MARCEL DEKKER INC, 1996.

183. Handbook of Lipids in Human Nutrition, Spiller G.A., CRC PRESS, 1996.

184. Supercritical Fluid Technology in Oil and Lipid Chemistry, King J.W., and List G., Ed., AOCS PRESS, 1996.

185. Lipoxygenase and Lipoxygenase Pathway Enzymes, Piazza G., Ed., AOCS PRESS, 1996.

186. Lipid Technologies and Applications, Gunstone F.D., and Padley F.B., Ed., MARCEL DEKKER, 1997.

187. Advances in Lipid Methodology – Four„ Christie W.W., Ed., OILY PRESS, 1997.

188. New Techniques and Applications in Lipid Analysis, McDonald R.E., Mossoba M.M., Eds, AOCS PRESS, 1997.

189. **Food Emulsifiers and their Applications,** Hasenhuettl G.L., Hartel R.W., Ed., CHAPMAN & HALL, 1997.

190. **Technology and Solvents for Extracting Oilseeds and Nonpetroleum Oils,** Wan P., Wakelyn P.J., Eds, AOCS PRESS, 1997.

191. **World Conference on Oilseed and Edible Oil Processing,** Kaseoglu S.S., Rhee K.C., Wilson R.F., Eds, AOCS PRESS, 1997.

192. **Alkyl Polyglycosides: Technology, Properties, and Applications,** Hill K., von Rybinski W., Stoll G., Eds, VCH, 1997.

193. **Natural Antioxidants – Chemistry, Health Effects, and Applications,** Shahidi F., Ed., AOCS PRESS, 1997.

194. **A Guide to Phospholipid Chemistry,** Hanahan D.J., OXFORD UNIVERSITY PRESS, 1997.

195. **Fats and Oils Handbook,** (Nahrungsfette und Öle), Bockisch M., AOCS PRESS, 1998.

196. **Lipid Analysis in Oils and Fats,** Hamilton R.J., Ed., BLACKIE, 1998.

197. **Food Lipids: Chemistry, Nutrition, and Biotechnology,** Akoh, C.C., Min D.B., Eds, AOCS PRESS, 1998.

198. **Choline, Phospholipids, Health, and Disease,** Zeisland S.H., Szuhaj B.F., Eds, AOCS PRESS, 1998.

199. **Lipids in Infant Nutrition,** Huang Y-S., Sinclair A.J., Eds, AOCS PRESS, 1998.

200. **Structural Modified Food Fats,** Christophe A., Ed., AOCS PRESS, 1998.

201. **Essential Fatty Acids and Eicosanoids: Invited Papers from the Fourth International Congress,** Riemersmaa R.A., Armsstrong R.A., Kelly R.W., Wilson R., Eds, AOCS PRESS, 1998.

202. **Biological Oxidants and Antioxidants: Molecules, Mechanisms, and Health Effects,** Packer L., Ong A.S.H., Eds, AOCS PRESS, 1998.

203. **Fatty Acids (Supplement to McCance and Widdowson's `The Composition of Foods')** Ministry of Agriculture, Fisheries & Food and Royal Society of Chemistry, UK, 1998.

204. *Trans* **FattyAcids in Human Nutrition,** Christie W.W., Sébédio J.L., Eds, THE OILY PRESS, 1998.

205. **Lipid Oxidation,** Frankel E.N., THE OILY PRESS, 1998.

206. **Fats and Oils: Formulating and Processing for Applications,** O'Brien R.D., TECHNOMIC, 1998.

207. **The Fats of Life,** Pond C.M., CAMBRIDGE UNIVERSITY PRESS, 1998

208. **Lipid Synthesis and Manufacture,** Gunstone F.D., Ed., SHEFFIELD ACADEMIC PRESS, 1999.

209. **Spectral Properties of Lipids,** Hamilton R.J., Cast, J., Eds, SHEFFIELD ACADEMIC PRESS, 1999.

210. **Oil World 2020,** ISTA MIELKE GMBH, 1999.

211. **Lipids in Nutrition and Health: A Reappraisal,** Gurr M.I., THE OILY PRESS, 1999.

212. **Recent Developments in the Synthesis of Fatty Acid Derivatives,** Knothe, G., Derksen, J.T.P., Eds, AOCS PRESS, 1999.

213. **Advances in Conjugated Linoleic Acid Research,** Vol.1, Yurawecz, M.P., Mossoba, M.M., Kramer, J.K.G., Pariza, M.W., Nelson, G., Eds, AOCS PRESS, 1999.

214. **Lipids in Health and Nutrition,** Tyman J.H.P., ROYAL SOCIETY OF CHEMISTRY, 1999.

215. **Phytochemicals and Phytopharmaceuticals,** Shahidi, F., Ho, C.T. Eds, AOCS PRESS, 1999.

216. **Frying of Food,** Boskou D., Elmadfa I., Eds, TECHNOMIC, 1999.

217. **Handbook of Dairy Foods and Nutrition,** 2nd Edn. Miller G.D., CRC PRESS, 1999.

218. **Emulsifiers,** Stauffer C.E., EAGAN PRESS, 1999.

219. **Vegetable Oils and Fats. Vol.1, Oils and Fats Handbook.** Rossell B.J., Ed., LEATHERHEAD FOOD RA, 1999.

220. **Fatty Acids in Foods and their Health Implications,** 2nd Edn. Ching Kuang Chow, Ed., MARCEL DEKKER, 2000.

221. **Handbook of Olive Oil,** Harwood J.L., Aparicio R., Eds, ASPEN, 2000.

222. **Lipid Glossary,** Gunstone, F.D., Herslöf, B.G., THE OILY PRESS, 2000.

223. **Edible Oil Processing,** Hamm W., Hamilton R.J., SHEFFIELD ACADEMIC PRESS, 2000.

224. **Introduction to Fats and Oils Technology,** 2nd Edn, Farr, W., O'Brien, R.D., Wan, P., Eds, AOCS PRESS, 2000.

225. **Physical Properties of Fats, Oils and Emulsifiers with Applications in Foods,** Widlak, N., Ed., AOCS PRESS, 2000.

226. **Fat Digestion and Absorption,** Ho Tan Hai, L., AOCS PRESS, 2000.

227. **Industrial Utilization of Surfactants,** Rosen, M.J. , Dahanayake, M., Rosen, AOCS PRESS, 2000.

228. **Animal Carcass Fats. Vol.2, Oils and Fats Handbook.** Rossell B.J., Ed., LEATHERHEAD FOOD RA, 2000.

Printed and bound by CPI Group (UK) Ltd, Croydon, CR0 4YY

03/10/2024

01040434-0001